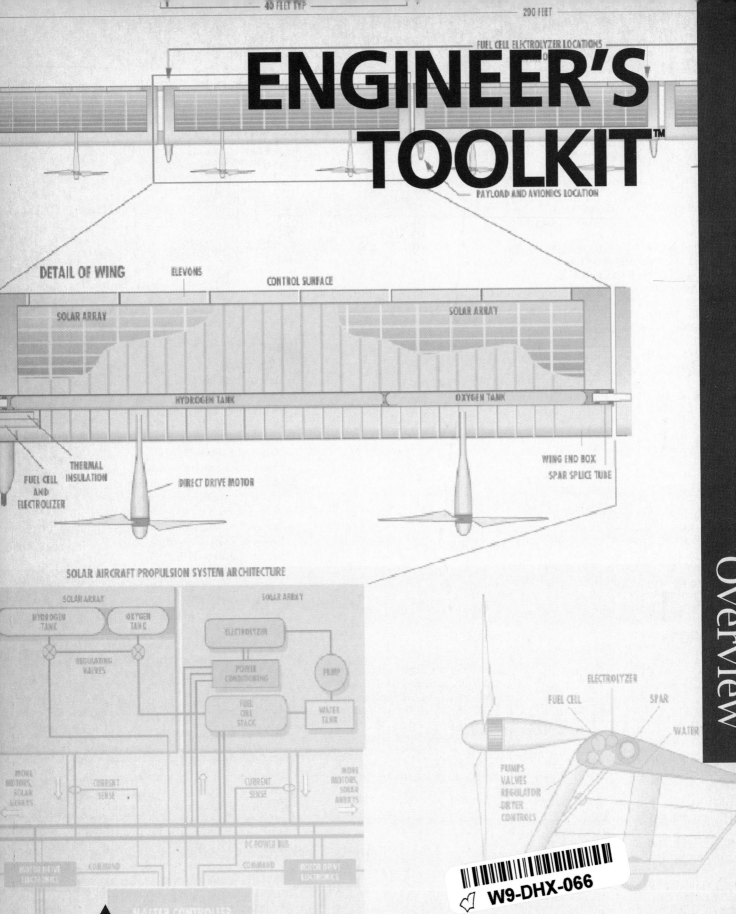

ENGINEER'S TOOLKIT™

Overview

40 FEET TYP — 290 FEET

FUEL CELL ELECTROLYZER LOCATIONS

PAYLOAD AND AVIONICS LOCATION

DETAIL OF WING

ELEVONS — CONTROL SURFACE

SOLAR ARRAY — SOLAR ARRAY

HYDROGEN TANK — OXYGEN TANK

FUEL CELL AND ELECTROLIZER — THERMAL INSULATION — DIRECT DRIVE MOTOR — WING END BOX / SPAR SPLICE TUBE

SOLAR AIRCRAFT PROPULSION SYSTEM ARCHITECTURE

SOLAR ARRAY — SOLAR ARRAY

HYDROGEN TANK — OXYGEN TANK — ELECTROLYZER

REGULATING VALVES — POWER CONDITIONING — PUMP

FUEL CELL STACK — WATER TANK

MORE MOTORS, SOLAR ARRAYS — CURRENT SENSE — CURRENT SENSE — MORE MOTORS, SOLAR ARRAYS

DC POWER BUS

COMMAND — COMMAND

MOTOR DRIVE ELECTRONICS — MOTOR DRIVE ELECTRONICS

MASTER CONTROLLER

ELECTROLYZER — FUEL CELL — SPAR — WATER

PUMPS VALVES REGULATOR DRYER CONTROLS

Addison-Wesley Publishing Company, Inc.

Menlo Park, California · Reading, Massachusetts · New York · Don Mills, Ontario
Wokingham, U.K. · Amsterdam · Bonn · Paris · Milan · Madrid · Sydney
Singapore · Tokyo · Seoul · Taipei · Mexico City · San Juan, Puerto Rico

Executive Editor: Dan Joraanstad
Acquisitions Editor: Denise Penrose
Marketing Manager: Mary Tudor
Developmental Editors: Deborah Craig,
Jeannine Drew, Kate Hoffman,
Shelly Langman
Assistant Editor: Nate McFadden
Senior Production Editor: Teri Holden
Production Editors: Jean Lake,
Gail Carrigan, Catherine Lewis
Supplements Production Editor: Teresa
Thomas
Photo Editor: Lisa Lougee
Copy Editors: Barbara Conway, Robert
Fiske
Proofreader: Holly McLean-Aldis
Marketing Coordinator: Anne Boyd
Cover Design: Yvo Riezebos
Text Design: Side by Side Studios
Technology Support: Craig Johnson
Composition: Side by Side Studios, Fog
Press, London Road Design, Progressive,
The Printed Page
Manufacturing Coordinator: Janet Weaver
Printing and Binding: R. R. Donnelley

Cover and Overview Photo Credits
Cover: Photo ©James Caccavo/Zuma
Images; background illustration
©Ian Worpole
Photo 1: Courtesy of NASA
Photo 2: Courtesy of Jet Propulsion
Laboratory
Photo 3: ©David Parker/SPL/Photo
Researchers, Inc.
Photo 4: ©Brownie Harris/The Stock
Market
Photo 5: Courtesy of M.E. Raichle, Wash.
Univ., St. Louis
Photo 6: ©Chuck O'Rear/Westlight
Photo 7: Courtesy of Lockheed. Photo by
Russ Underwood.
Photo 8: ©Roger Ressmeyer/Starlight
Photo 9: ©George Haling/Photo
Researchers, Inc.
Photo 10: Courtesy of Keith Wood,
Promega, Inc.

ISBN: 0-8053-6335-1

Addison-Wesley Publishing Company, Inc.
2725 Sand Hill Road
Menlo Park, CA 94025

COVER STORY

Pictured on the cover of The Engineer's
Toolkit is the Pathfinder—a "solar-powered
flying wing" designed for low-speed, high-
altitude flight. With a wing span compara-
ble to a Boeing 737, it weighs in at just 400
pounds and has no rudders, no fins, no tail,
and no pilot! The Pathfinder is one of a
series of solar planes developed and built
by Dr. Paul MacCready and his team of
engineers at AeroVironment Inc., in Simi
Valley, California. Engineers at the
Lawrence Livermore National Laboratory in
Livermore, California, designed, engineered, and continue to administer the
Pathfinder solar plane. This laboratory also
is designing the next iteration of solar
planes, the Helios (plans for which appear
behind the photograph of the Pathfinder).
With the Helios, engineers hope to come
even closer to realizing the dream of "eter-
nal flight"; it will include an on-board
energy storage system that can provide the
energy needed during night flight.

As with most contemporary engineering projects, designing solar planes requires
the efforts of engineers from a variety of
disciplines—aeronautical, computer, electrical, environmental, and mechanical, to
name a few. Still other teams of engineers
are needed to design on-board equipment
to support specific missions, such as moni-
toring dangerous weather systems or tracking the release of toxins into the atmosphere.

Contemporary design examples such as
these are presented throughout The Engineer's Toolkit, highlighting the interdisciplinary teamwork that characterizes engineering today.

Tools for a New Curriculum

The Engineer's Toolkit is not a conventional textbook. It was inspired by the needs of instructors like you, who are engaged in developing a new curriculum in introductory engineering courses. They are searching for new ways to prepare, motivate, and engage first-year students. They want to create a link for their students between the prerequisite math and science courses and the wide range of engineering disciplines that build on that knowledge. These instructors also want to ensure that their students master the skills of team-building, communications, and computer use—skills they need to solve problems successfully in subsequent courses and in the real world of work. You and your colleagues are also experimenting with hands-on design projects so students understand that design is a process and that, fundamentally, engineering means solving problems.

Universities and colleges are responding in unique ways to the changing landscape of introductory engineering. This very uniqueness creates a new challenge when you are searching for the right text to support your unique course. The Engineer's Toolkit takes on that challenge. You choose from a rich set of course materials that introduce fundamental concepts of engineering and teach essential skills and tools. Each tool is presented as a single module. You determine which modules will best satisfy your course goals, and Addison-Wesley binds those modules into the exact book your students need.

Especially written and designed for The Engineer's Toolkit, the modules present a consistent teaching methodology adapted from the work of Delores Etter, author of the spreadsheet and Fortran Toolkit modules, as well as Structured Fortran 77 for Scientists and Engineers. Each of the Toolkit authors has applied

ENGINEER'S TOOLKIT

A FIRST COURSE IN ENGINEERING

AN ADDISON-WESLEY SELECT EDITION

Dr. Etter's five-step problem-solving process to a wide variety of programming languages and application programs. A consistent approach, style, level, and tone means you and your students don't have to switch gears every time you begin to teach or learn a new tool.

Here are the six key pedagogical features you'll find in the Toolkit modules that teach programming languages or software tools:

◆ **The five-step problem-solving process** is explained and illustrated in terms of the particular language or software tool being taught. It is then used throughout the module in applications, numbered examples, and end-of-chapter exercises or problems.

◆ **Applications** based on the Ten Great Engineering Achievements and representing a wide variety of engineering disciplines demonstrate the five-step problem-solving process.

◆ **"What If?" problems** immediately follow applications in the software tools modules, asking students to modify assumptions, data, or variables in the application and to solve the new problems that result.

◆ **Numbered examples** demonstrate key elements of a language or application program by providing fully worked-out solutions.

◆ **"Try It!" exercises** test students' knowledge of sections within a chapter and frequently require work at the computer.

◆ **End-of-chapter material** includes summaries of essential points, a key word list, and a set of exercises or problems that gradually increase in complexity.
These pedagogical features are also described from a student's point of view in the section "The Toolkit Methodology."

How To Design Your Custom Textbook

A sample course goal: To introduce engineering, teach a programming language, word processing and CAD techniques.

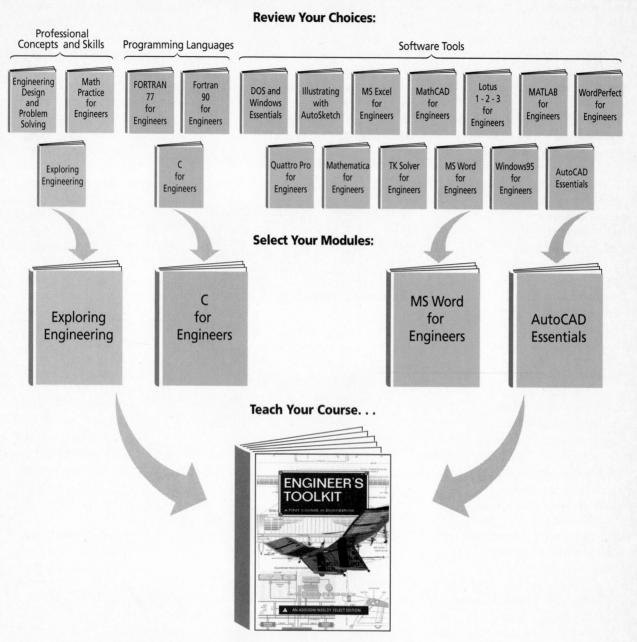

Review Your Choices:

Professional Concepts and Skills — Programming Languages — Software Tools

Engineering Design and Problem Solving | Math Practice for Engineers | FORTRAN 77 for Engineers | Fortran 90 for Engineers | DOS and Windows Essentials | Illustrating with AutoSketch | MS Excel for Engineers | MathCAD for Engineers | Lotus 1-2-3 for Engineers | MATLAB for Engineers | WordPerfect for Engineers

Exploring Engineering | C for Engineers | Quattro Pro for Engineers | Mathematica for Engineers | TK Solver for Engineers | MS Word for Engineers | Windows95 for Engineers | AutoCAD Essentials

Select Your Modules:

Exploring Engineering | C for Engineers | MS Word for Engineers | AutoCAD Essentials

Teach Your Course...

ENGINEER'S TOOLKIT
A FIRST COURSE IN ENGINEERING

AN ADDISON-WESLEY SELECT EDITION

With Your Custom Textbook!

DESIGN YOUR OWN TOOLKIT

The *Toolkit* menu consists of three modules that teach programming languages, thirteen modules that teach software tools, and three modules that focus on core engineering concepts and skills—*Engineering Design and Problem Solving, Exploring Engineering,* and *Math Practice for Engineering,* and *Math Practice for Engi-neers.* You can mix and match as your course demands.

In a course that combines an overview of engineering disciplines with essential computer applications and the basic design process, you might, for example, combine *Exploring Engineering, Engineering Design and Problem Solving, Quattro Pro for Engineers,* and *Microsoft Word for Engineers* for a book of about 460 pages.

If you plan to teach programming and introduce your students to design and basic CAD techniques, you might combine *Engineering Design and Problem Solving, C for Engineers,* and *AutoCAD Essentials* for a book of about 550 pages.

The *Toolkit* modules are also available separately, and each custom text includes this Overview, which introduces instructors and students to the *Toolkit* methodology.

FUTURE MODULES

Each year Addison-Wesley will release additional modules to keep pace with the development in introductory engineering courses. We welcome your suggestions for future modules in *The Engineer's Toolkit*. Please correspond with us at the Internet address

toolkit@aw.com

or by regular mail at

Toolkit
Addison-Wesley Publishing
Company, Inc.
2725 Sand Hill Road
Menlo Park, CA 94025

Toolkit information is also available on the World Wide Web at

http://www.aw.com/cseng/toolkit/

ELECTRONIC SUPPLEMENTS AND SUPPORT MATERIAL

Addison-Wesley offers a range of support materials for *The Engineer's Toolkit*, both printed and electronic. We welcome your comments on the effectiveness of these materials and your suggestions for additional supplements.

Supplements via the Internet A selection of supplementary material for *The Engineer's Toolkit*, including transparency masters, is available on the Internet via anonymous FTP. The URL for accessing this material is ftp://aw.com/cseng/toolkit.

Instructor's Guides Instructor's Guides for the modules you choose will be provided upon adoption of *The Engineer's Toolkit*.

Each Instructor's Guide opens with an overview of the module subject matter, suggested methods of instruction, comments on tests and quizzes, and a general discussion of software version and platform issues (if applicable).

Each Instructor's Guide also describes how to order the module topics to support various syllabi.

A chapter-by-chapter section presents teaching strategies, points to emphasize and special challenges for each chapter, solutions to the end-of-chapter problems, and additional problems and solutions.

Student Data Files Student Data Files, including the data, applications, and program files that support end-of-chapter problems in specific modules, will be available in the following ways:

- ◆ on the disk that includes the Instructor's Guide
- ◆ via our FTP site: ftp://aw.com/cseng/toolkit/igs/
- ◆ via our Toolkit home page on the World Wide Web: http://www.aw.com/cseng/toolkit/

Modules with Data Files include *AutoCAD Essentials, Illustrating with AutoSketch, FORTRAN 77 for Engineers, Fortran 90 for Engineers, C for Engineers, Quattro Pro for Engineers, Lotus 1-2-3 for Engineers, Microsoft Excel for Engineers, TK Solver for Engineers,* and *Mathematica for Engineers.*

Errata for Published Modules: Errata notices for published modules will be available online at: http://www.aw.com/cseng/toolkit/

FIRST ASSIGNMENT

The next section of this Overview is directed to your students. It explains the teaching and learning strategies adopted by our authors throughout the *Toolkit*. The final section introduces students to the Ten Great Engineering Achievements. We invite you to read on and hope you will assign this section to your students early in your course.

THE ENGINEER'S TOOLKIT MODULES

	Title	Author
Professional Concepts and Skills	Engineering Design and Problem Solving	Steve Howell
	Exploring Engineering	Joe King
	Math Practice for Engineers	Joe King
Software Tools	AutoCAD Essentials	Melton Miller
	DOS and Windows Essentials	Gerald Lemay
	Illustrating with AutoSketch	Gordon Snyder
	Lotus 1-2-3 for Engineers	Delores Etter
	MathCAD for Engineers	Joe King
	Mathematica for Engineers	Henry Shapiro
	MATLAB for Engineers	Joe King
	Microsoft Excel for Engineers	Delores Etter
	Microsoft Word for Engineers	Sheryl Sorby
	Quattro Pro for Engineers	Delores Etter
	TK Solver for Engineers	Robert J. Ferguson
	Windows95 for Engineers	Gordon Snyder
	WordPerfect for Engineers	Sheryl Sorby
Programming Languages	C for Engineers	Kenneth Collier
	FORTRAN 77 for Engineers	Delores Etter
	Fortran 90 for Engineers	Delores Etter

The Toolkit Methodology

Welcome to *The Engineer's Toolkit!* This book has been especially created to support your work in what is probably your first course in Engineering. Unlike other textbooks you have studied, *The Engineer's Toolkit* was customized by your instructor to include the exact material you need, and only the material you need. In essence, *The Engineer's Toolkit* is a collection of modules that teach engineering concepts and skills, software tools, and programming languages. Your instructor has selected the appropriate modules for your course, and Addison-Wesley has bound those modules into this custom book.

Introduction. Each application is fully described and explained so that you have sufficient information to complete step 1.

A GENERAL PROCESS FOR SOLVING PROBLEMS

A key feature of *The Engineer's Toolkit* is its emphasis on developing problem-solving skills. Problem solving is one of the foundations of all engineering activity. In *The Engineer's Toolkit* you'll find a five-step method for solving the problems given in each module. Some engineers will tell you they use a nine-step process; others can condense their process down to four. There's nothing magic about the number, but you will find that learning and following a consistent method for solving problems will make you an efficient student and a promising graduate. Each application program or programming language module builds on this general problem-solving method:

1. Define the problem.
2. Gather information.
3. Generate and evaluate potential solutions.
4. Refine and implement the solution.
5. Verify the solution through testing.

---

66   QUATTRO PRO FOR ENGINEERS

and choosing it. An icon will be on this page for each graph book. Choose the icon of the graph you want to rename with and select Properties|Current Object. The Name dialog box in which you may type the new name of the graph and then c

Saving and printing graphs is similar to saving and print sheets. The graph is automatically saved when you save the file. You can print the graph either by itself or with the spre print the graph by itself, select the graph using Graph|Edi choose File|Print. If you have placed the graph as a floating g spreadsheet, it is printed when you print the spreadsheet.

 Try It   Change the name of the GRAPH1 graph in the FILTER1 spreads NALIN. Then print the graph.

Application 1   QUALITY CONTROL

Manufacturing Engineering
In a manufacturing or assembly plant, quality control receives tion. One of the key responsibilities of a quality control engine lect accurate data on the quality of the product being manufa data can be used to identify the problem areas in the assemb the materials being used in the product.

Circuit Board Defects
In this application, information collected over a one-year perio specify both the type of defects and the number of defects det assembly of printed circuit boards. These defects have been four categories: board errors, chip errors, processing errors, tion errors. Board errors are typically caused by defects in the of the printed circuits. Chip errors are caused by defective in cuit (IC) chips that are added to the board; these IC chips incl chips, microprocessor chips, and digital filter chips. Processin typically caused by errors in inserting the chips in the board; is often done by manufacturing robots, and the robot progra be incorrect, or the chips being inserted can be packaged in order. Connection errors are solder errors that occur when thi through the solder machine; these errors can be caused by board or an incorrect solder temperature.

Spreadsheet for a Quality Analysis Report
You want to develop a spreadsheet that summarizes the qua data that has been collected each month for a year. This data number of defects in each of the four categories of defects dis summary report should compute totals and percentages fo year and defect totals for each quarter. Later in this chapter the data in the spreadsheet to generate pie, bar, and line grap

From *Quattro Pro® for Engineers*

---

184   C FOR ENGINEERS

```c
void init_array(struct sample_type array[MAX_RAINFALLS][MAX_SITES])
{
    int row, col;        /* Loop control variables */

    for (row=0; row < MAX_RAINFALLS; ++row)
        for (col=0; col < MAX_SITES; ++col){
            array[row][col].date.day = 0;
            array[row][col].date.month = 0;
            array[row][col].date.year = 0;
            array[row][col].time.hour = 0;
            array[row][col].time.minutes = 0;
            array[row][col].h_concentration = 0.0;
            array[row][col].ph_level = 0.0;
        }
}
```

. . . .

Carefully examine the definition of `init_array()`. Notice that the parameter array is declared as a two-dimensional array of structures. In the body of this function access is made to elements within `array` using the familiar subscript notation. Once a particular element has been accessed in this manner, the fields of that structure are accessed using the dot operator. Whenever the accessed field is itself a structure, its fields are accessed using a second dot operator, as in the statement

```c
array[row][col].date.month = 0;
```

Because the dot is an operator, it can be combined with other operators in this way.

There are many powerful ways to use structures in a program, and this section provides you with only an introduction. In the following application you will learn how structures can be used to extend the mathematical power of *C* to include complex arithmetic.

Application 1   FREQUENCY DISTRIBUTION GRAPHING

Industrial Engineering
Quality-control engineers monitor the quality of an automated production line by tracking the number of defective parts coming off the line within a particular period. If the frequency of defective parts rises dramatically for a given period, the engineer is alerted that a problem exists and can take action to fix the problem. Such frequencies can be depicted using a bar graph such as the one shown in Figure 7-5. In this graph, the horizontal axis represents the number of defects detected and the vertical axis represents data collection periods.

From *C for Engineers*

CHAPTER 4 ENGINEERING GRAPHS 67

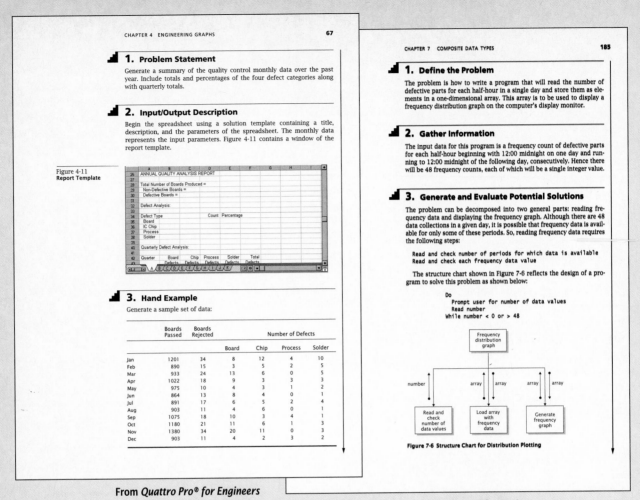

1. Problem Statement

Generate a summary of the quality control monthly data over the past year. Include totals and percentages of the four defect categories along with quarterly totals.

2. Input/Output Description

Begin the spreadsheet using a solution template containing a title, description, and the parameters of the spreadsheet. The monthly data represents the input parameters. Figure 4-11 contains a window of the report template.

Figure 4-11
Report Template

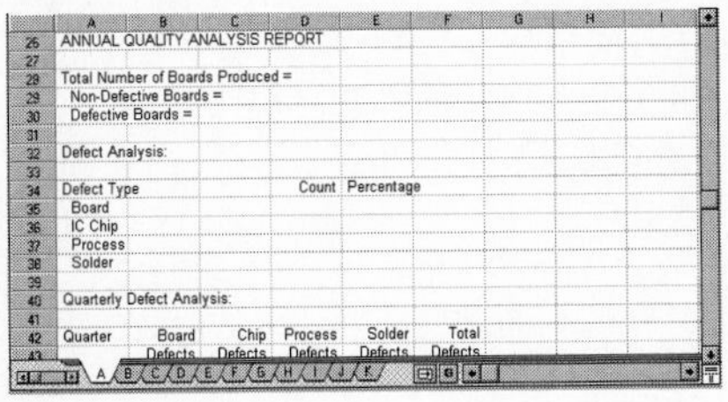

3. Hand Example

Generate a sample set of data:

	Boards Passed	Boards Rejected	Number of Defects			
			Board	Chip	Process	Solder
Jan	1201	34	8	12	4	10
Feb	890	15	3	5	2	5
Mar	933	24	13	6	0	5
Apr	1022	18	9	3	3	3
May	975	10	4	3	1	2
Jun	864	13	8	4	0	1
Jul	891	17	6	5	2	4
Aug	903	11	4	6	0	1
Sep	1075	18	10	3	4	1
Oct	1180	21	11	6	1	3
Nov	1380	34	20	11	0	3
Dec	903	11	4	2	3	2

From *Quattro Pro® for Engineers*

CHAPTER 7 COMPOSITE DATA TYPES 185

1. Define the Problem

The problem is how to write a program that will read the number of defective parts for each half-hour in a single day and store them as elements in a one-dimensional array. This array is to be used to display a frequency distribution graph on the computer's display monitor.

2. Gather Information

The input data for this program is a frequency count of defective parts for each half-hour beginning with 12:00 midnight on one day and running to 12:00 midnight of the following day, consecutively. Hence there will be 48 frequency counts, each of which will be a single integer value.

3. Generate and Evaluate Potential Solutions

The problem can be decomposed into two general parts: reading frequency data and displaying the frequency graph. Although there are 48 data collections in a given day, it is possible that frequency data is available for only some of these periods. So, reading frequency data requires the following steps:

```
Read and check number of periods for which data is available
Read and check each frequency data value
```

The structure chart shown in Figure 7-6 reflects the design of a program to solve this problem as shown below:

```
Do
  Prompt user for number of data values
  Read number
While number < 0 or > 48
```

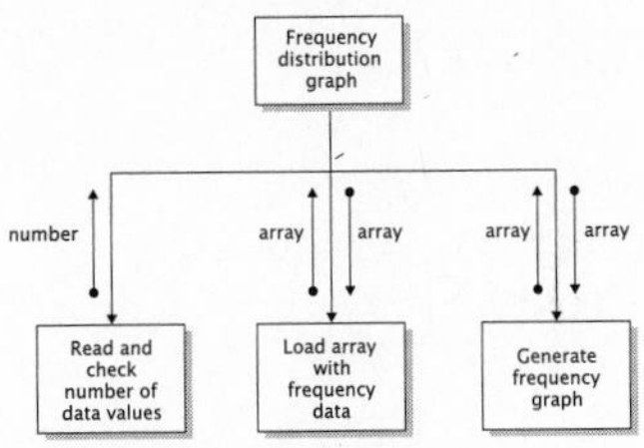

Figure 7-6 Structure Chart for Distribution Plotting

From *C for Engineers*

Step 1. In both applications you are asked to **define the problem.** The introductory description gives lots of clues to help you.

Step 2. This step asks you to **gather information** that you need to propose a solution. In these applications you need to prepare the data that will be used to generate quality analysis reports.

Step 3. You are now ready to **generate and evaluate potential solutions.** In C you create a structure chart to illustrate the design of a program that can solve the problem. In Quattro Pro, the hand example shows how the gathered data will be used to determine the algorithm in step 4.

A SPECIFIC FIVE-STEP PROCESS

Each module adapts this general method and refines it according to the kinds of problems solved by the tool or language being taught. Chapter 1 of each programming language and application program module describes the specific five-step process used in that module.

As you work through *The Engineer's Toolkit*, you'll find that this consistent approach makes it easier to solve new problems. For instance, step 1 of the five-step problem-solving process calls for the same kind of thinking process whether you are using a programming language like Fortran 90, a computer-aided design package like AutoCAD, or an equation solver like MATLAB.

We illustrate the five-step problem-solving process with a pair of applications from *The Engineer's Toolkit.* Both of the examples presented here deal with the collection and tracking of data related to quality control. Each has been fully worked out using the five-step process. Follow these steps to see how easy it is to learn this problem-solving process.

APPLICATIONS

Step 4. In this step you write a C program based on the structure chart and algorithms developed in the previous steps. You will now **refine and implement the solution.** With Quattro Pro, this step means developing the formulas that will be used to compute the values listed in the summary report.

Step 5. In the final step you **verify the solution through testing.** In C this involves entering a variety of values (value testing) to confirm that the program generates valid output. In Quattro Pro the spreadsheet is tested with several sets of data to verify the accuracy of the computations. Accuracy is confirmed by comparing the spreadsheet calculations to the values determined in the hand example.

These two applications are among hundreds in which *Toolkit* authors demonstrate the five-step problem-solving method. As you gain experience using this method with various software tools and languages, you'll find you can approach new problems with confidence, and you'll begin to identify the appropriate tool or language for the problem at hand. Learning to choose the right tool for a specific engineering problem is an important part of your education.

Many of the applications in *The Engineer's Toolkit* are based on the Ten Great Engineering Achievements chosen by the National Academy of Engineering to celebrate its silver anniversary in 1989. Studying these applications will help you see the kinds of problems faced by engineers from different disciplines and better understand how large problems are broken into smaller solvable problems. This Overview concludes with a description of the Ten Great Engineering Achievements.

4. Algorithm Development

The spreadsheet now contains everything except the formulas for computing the summary information for the report. Develop the formulas in the order needed to compute the values. Also, try to minimize the number of computations. For example, since you generate error sums by quarters, add the quarterly sums to get yearly sums instead of adding all the monthly sums to get yearly sums. It would be good practice to verify each of these formulas by referring to Figures 4-10 and 4-11.

D29	@SUM(D11..D22)	Total non-defective boards
D30	@SUM(E11..E22)	Total defective boards
E28	+D29+D30	Total boards
E29	+D29/E28	Percent non-defective boards
E30	Copy from E29	Percent defective boards
B44	@SUM(G11..G13)	Quarter I board defects
B45	@SUM(G14..G16)	Quarter II board defects

5. Testing

An important part of developing a spreadsheet is testing it with several sets of data to verify the accuracy of the computations. Using the sample set of data from the hand example, you can easily check the accuracy of the spreadsheet calculations by comparing them to the hand example.

You should make minor changes in this data and check the report to be sure that corresponding changes occur in the report summary. In this report you want to be sure that the report would be generated correctly if there were no errors in one of the categories. Be sure to change the corresponding sums of boards rejected. The corresponding report generated, shown in Figure 4-12, shows that there were no process defects during any of the four quarters.

Figure 4-12
Report with No Processing Defects

ANNUAL QUALITY ANALYSIS REPORT

Total Number of Boards Produced =		12,423
Non-Defective Boards =	12,217	98.34%
Defective Boards =	206	1.66%

Defect Analysis:

Defect Type	Count	Percentage
Board	100	48.54%
IC Chip	66	32.04%
Process	0	0.00%
Solder	40	19.42%

Quarterly Defect Analysis:

Quarter	Board Defects	Chip Defects	Process Defects	Solder Defects	Total Defects
I	24	23	0	20	67
II	21	10	0	6	37
III	20	14	0	6	40
IV	35	19	0	8	62

From *Quattro Pro® for Engineers*

4. Refine and Implement a Solution

The structure chart and algorithms developed in the previous section are implemented as the program in Example 7-10.

5. Verify and Test the Solution

To properly test this program, you should enter a variety of values for first input, including 0, 48, values below 0, values above 48, and valid values. Selecting values along the boundaries is known as boundary value testing. You should do the same for the actual frequency values. Given the input values −3 (error) 55 (error) 10−9 (error) 15 16 13 14 20 9 96 (error) 19 14 19 20, the output of this program is

```
 --------------------------------------------------
  1 | ***************
  2 | ****************
  3 | *************
  4 | **************
  5 | ********************
  6 | *********
  7 | ********************
  8 | ***************
  9 | **************
 10 | *********************
     --- 5---10---15---20---25---30---35---40---45---50
```

This chapter introduced you to one-, two- and three-dimensional arrays, as well as structures. Arrays and structures offer a cohesive means of storing composite collections of data. Arrays are used for grouping items of the same type and meaning, whereas structures allow you to group related items with different types and meanings. You learned how to declare and initialize arrays and manipulate the elements in an array. You also learned how to define a structure, declare structure variables, and manipulate the fields in a structure. Finally, you learned that these composite data types can be combined to allow you to create arrays of structures, structures containing arrays, and structures containing other structures.

From *C for Engineers*

68 AUTOCAD ESSENTIALS

5. Draw, Print or Plot, and Save the Drawing File

Draw the schematic as shown. The lenses are arcs. You draw an arc by first drawing a circle and intersecting it with a line. Then you trim away the unwanted part of the circle. The other face of the lens is a mirror image of the first face. After you have drawn one lens, the remaining drawing is quite straightforward. Do not attempt to add the text.

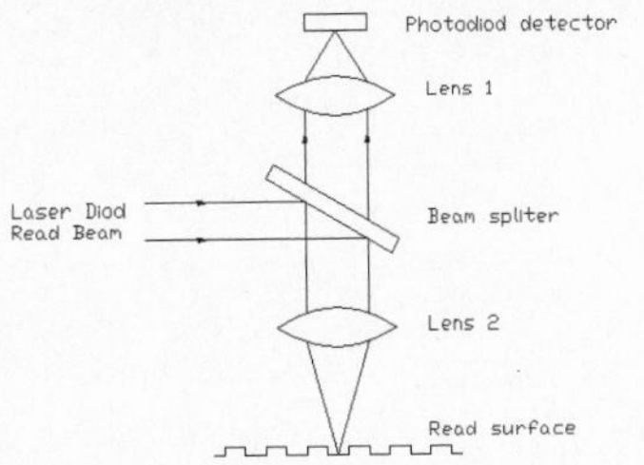

Photodiod detector
Lens 1
Laser Diod
Read Beam
Beam spliter
Lens 2
Read surface

Figure 3-29
Optical Disk Pickup Schematic

What If What if the drawing is too long vertically? You can shorten the drawing by shortening the light rays between lens 1 and the reflecting mirror. Using a horizontal line as a cutting edge, trim the lines between the mirror and the lens. Finally, move the lens and the photodiode detector down. Use the top surface of the beam splitter as a boundary edge, and extend the vertical lines to intersect the mirror. You can use the same technique to shorten the lines between lens 2 and the beam splitter.

SUMMARY

In this chapter you've added several new editing tools to your repertoire. You can copy, move, and mirror objects. The Offset command is especially convenient if you need parallel or concentric copies of objects. You have learned how to extend and trim objects. Layers are a powerful tool you can use to organize a drawing. Linetypes and colors add a professional quality to the result. The Chprop command is useful when you need to change the properties of selected objects. You can correct mistakes by using the Undo and Redo commands. Finally, the Redraw command permits you to refresh your screen, and the Zoom and Pan commands help you when you need to see details close up.

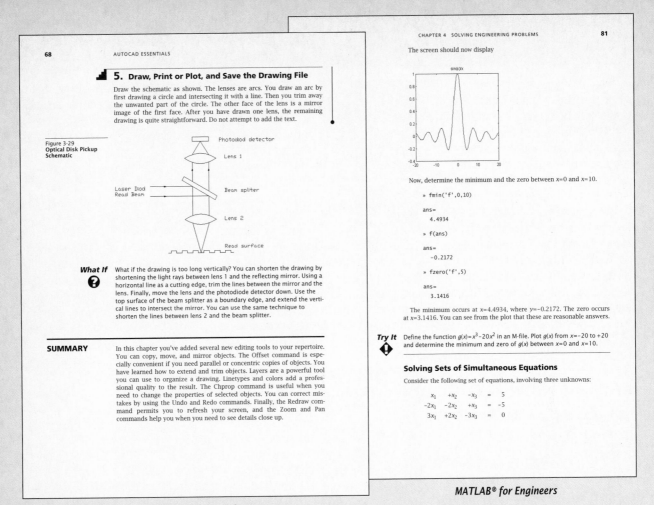

AutoCAD® Essentials

CHAPTER 4 SOLVING ENGINEERING PROBLEMS 81

The screen should now display

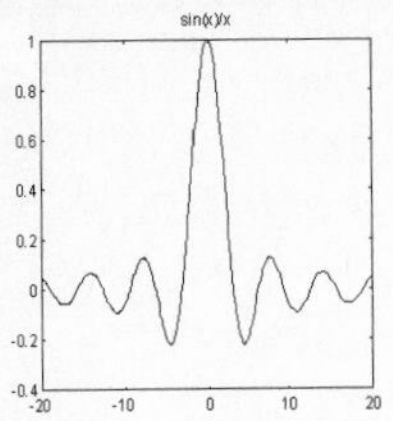

sin(x)/x

Now, determine the minimum and the zero between x=0 and x=10.

```
» fmin('f',0,10)

ans=
   4.4934

» f(ans)

ans=
   -0.2172

» fzero('f',5)

ans=
   3.1416
```

The minimum occurs at x=4.4934, where y=−0.2172. The zero occurs at x=3.1416. You can see from the plot that these are reasonable answers.

Try It Define the function $g(x)=x^3-20x^2$ in an M-file. Plot $g(x)$ from x=−20 to +20 and determine the minimum and zero of $g(x)$ between x=0 and x=10.

Solving Sets of Simultaneous Equations

Consider the following set of equations, involving three unknowns:

$$\begin{aligned} x_1 + x_2 - x_3 &= 5 \\ -2x_1 - 2x_2 + x_3 &= -5 \\ 3x_1 + 2x_2 - 3x_3 &= 0 \end{aligned}$$

MATLAB® for Engineers

ENGINEERING DISCIPLINES

In *AutoCAD Essentials*, the author presents applications from mechanical and materials engineering, civil and structural engineering, electrical and electronics engineering, and optical engineering. *MATLAB for Engineers* includes applications from electrical and computer engineering. You can find the listing of application problems in Chapter 1 of each software tool or programming language module. The table below lists several applications from the FORTRAN 77 module.

Applications	ACROSS THE DISCIPLINES	Chapter
Stride Estimation	Mechanical Engineering	2
Light Pipes	Optical Engineering	3
Sonar Signals	Acoustical Engineering	4
Wind Tunnels	Aerospace Engineering	5
Oil Well Production	Petroleum Engineering	6
Simulation Data	Electrical Engineering	7

"What If" Problems

These problems immediately follow an application. Often you are asked to test the model developed in the application with different data or different assumptions. Working through these problems ensures you fully understand the application.

"Try It!" Exercises

Each chapter contains several "Try It!" exercises. A "Try It!" is a short set of exercises that tests your understanding of the material. These exercises increase in complexity over the course of each module, and if you try out each one, you'll find you'll master the material faster than you expected.

NUMBERED EXAMPLES

Numbered examples appear in all the programming language and some software tool modules. These examples are designed to illustrate specific features of the language or software. Working through these examples is essential, especially those that offer two solutions and a discussion of the differences between them.

EXERCISES/PROBLEMS

Every module includes end-of-chapter exercises or problems that increase in difficulty and test your knowledge of the chapter material. Make sure you try your hand at the excercises that require you to use the five-step method to find solutions.

The *Toolkit* Team

As you consider the Ten Great Engineering Achievements explored at the end of this Overview, you'll notice that many require contributions from several different engineering disciplines. Today, significant projects can only be accomplished by teams of professionals. And that's true of *the Engineer's Toolkit,* too. The team that has helped to create the *Toolkit* includes not only the authors but also the focus group participants who honed and directed the *Toolkit* concept and the reviewers who helped develop the individual manuscripts.

TOOLKIT AUTHORS

Delores Etter, author of five modules, *FORTRAN 77 for Engineers, Fortran 90 for Engineers, Lotus 1-2-3 for Engineers, QuattroPro for Engineers, and Microsoft Excel for Engineers,* is a professor of electrical and computer engineering at the University of Colorado at Boulder. Dr. Etter has helped shape *The Engineer's Toolkit* from its initial conception, contributing the five-step problem-solving process and key pedagogical features that were successfully tested in her earlier Addison-Wesley texts, such as *Structured FORTRAN 77 for Engineers and Scientists, Lotus 123: A Software Tool for Engineers,* and *Quattro Pro: A Software Tool for Engineers.*

Ken Collier, Assistant Professor of Computer Science in the Department of Computer Science and Engineering at Northern Arizona University in Flagstaff, is the author of *C for Engineers.* Professor Collier teaches courses in C, C++, software engineering, and engineering design. His areas of research include software engineering, software design methodologies, and artificial intelligence.

R. J. Ferguson, the author of *TK Solver for Engineers,* is a professor of mechanical engineering at the Royal Military College of Canada. He teaches courses in stress analysis and computer-aided design. Other publications by Professor Ferguson include texts in the fields of fracture mechanics, noncircular gearing, vehicle transmissions, and engineering education.

Steve Howell, Associate Professor of Engineering in the mechanical engineering department at Northern Arizona University in Flagstaff, Arizona, is the author of *Engineering Design and Problem Solving.* He teaches a course titled Introduction to Engineering Design and Graphics. Professor Howell's areas of research are computer-aided design and manufacturing, thermodynamics, and heat transfer.

Joe King, the author of *Math-CAD for Engineers, MATLAB for Engineers, Exploring Engineering,* and *Math Practice for Engineers,* is an associate professor of electrical engineering at the University of the Pacific in Stockton, California. He teaches courses in electrical engineering, advanced digital design, local area networks, neural networks, machine vision, C++, and microprocessor applications. He conducts research in the areas of neural networks and microprocessor applications.

Gerald Lemay, Professor of Electrical and Computer Engineering at the University of Massachusetts, Dartmouth, is the author of *DOS and Windows Essentials.* He teaches the Science of Engineering for honors students and Computer Tools for Engineers. Professor Lemay does research in renewable energy.

Melton Miller, author of *Auto-CAD Essentials,* is an associate professor of civil engineering and assistant dean of the College of Engineering at the University of Massachusetts, Amherst. He teaches courses in Pascal, Fortran, Lotus, MathCAD, and AutoCAD. He also teaches courses in the the design of reinforced concrete structures.

Henry Shapiro is the author of *Mathematica for Engineers.* He is an associate professor of computer science at the University of New Mexico, where he teaches courses in computer programming and mathematical foundations of computer science. Professor Shapiro conducts research in the area of algorithm design. He is also active in curriculum development and accreditation of computer science programs.

Sheryl Sorby is the author of *WordPerfect for Engineers* and *Microsoft Word for Engineers.* She is an assistant professor of civil and environmental engineering at Michigan Technological University in Houghton, where she teaches courses in freshman engineering and computer skills. Professor Sorby conducts research in structural engineering.

Gordon Snyder, author of *Illustrating with AutoSketch* and *Windows 95 Essentials,* is an associate professor and department co-chair of the departments of electronics systems engineering, computer systems engineering, and laser electro-optics technology at Springfield Technical Community College in Springfield, Massachusetts. He teaches a course titled Introduction to Computer-Aided Engineering Technology.

REVIEWERS AND FOCUS GROUP PARTICIPANTS

Instructors throughout the country attended focus groups to help us identify key trends in engineering education. Over 100 reviewers contributed to the development of manuscripts for *The Engineer's Toolkit.* We gratefully acknowledge all their contributions.

Teresa Adams, University of Wisconsin-Madison · Anjum Ali, Mercer University · Abbas Aminmansour, Pennsylvania State University · A. Adnan Aswad, University of Michigan-Dearborn · Stormy Attaway, Boston University · James E. Bailey, Arizona State University · Betty Barr, University of Houston · O. Barron, University of Tennessee at Martin · Charles Beach, Florida Institute of Technology · Bill Beckwith, Clemson University · Charlotte Behm, Mission College · Lynn Bellamy, Arizona State University · Bhushan L. Bhatt, Oakland University · Steve Borgelt, University of Missouri · Joseph M. Bradley, United States Navy · Robert Brannock, Macon College · Gus Brar, Delaware Community College · David Bricker, Oakland University · Matthew Calame, Cosumnes River College · Christopher Carroll, University of Minnesota-Duluth · R. J. Coleman, University of North Carolina-Charlotte · James Collier, Virginia Polytechnic Institute and State University · Kenneth Collier, Northern Arizona University · Tom Cook, Mercer University · John Barrett Crittenden, Virginia Polytechnic Institute and State University · Barry Crittendon, Virginia Polytechnic Institute and State University · James Cunningham, Embry-Riddle Aeronautical University · Janek Dave, University of Cincinnati · Tim David, Leeds University · Al Day, Iowa State University · Jack Deacon, University of Kentucky · Bruce Dewey, University of Wyoming · Julie Ellis, University of Southern Maine · Beth Eschenbach, Humboldt State University · David Fletcher, University of the Pacific · Wallace Fowler, University of Texas-Austin · Jim Freeman, San Jose State University · Ben Friedman, Wolfram Research, Inc. · Paul Funk, University of Evansville · David Gallagher, Catholic University of America · Byron Garry, South Dakota State University · Larry Genalo, Iowa State University · Alan Genz, Washington State University · Johannes Gessler, Colorado State University · John J. Gilheany, Catholic University of America · Oscar R. Gonzalez, Old Dominion University · Yolanda Guran, Oregon Institute of Technology · Nicolas Haddad, New Mexico Institute of Mining and Technology · John Hakola, Hofstra University · Fred Hart, Worcester Polytechnic University · David Hata, Intel Corporation · Frank Hatfield, Michigan State University · Jim D. Helmert, Eastern Oklahoma State University · Herman W. Hill, Ohio University · Deidre A. Hirschfeld, Virginia Polytechnic Institute and State University · Margaret Hoft, University of Michigan-Dearborn · Henry Horwitz, Dutchess Community College · Peter K. Imbrie, Texas A & M University · Jeanine Ingber, University of New Mexico · Scott Iverson, University of Washington · Denise Jackson, University of Tennessee · Gerald S. Jakubowski, Loyola Marymount University · Scott James, GMI Engineering and Management Institute · Sundaresan Jayaraman, Georgia Institute of Technology · Johndan Johnson-Eilola, Purdue University · Douglas Jones, George Washington University · John Kelly, Arizona State University · Robert Knighton, University of the Pacific · Bill Koffke, Villanova University · Celal N. Kostem, Lehigh University · George V. Krestas, De Anza College · Tom Kurfess, Carnegie Mellon University · Ronald E. Lacey, Texas A & M University · Frank Lee, Bellevue Community College · Gerald J. Lemay, University of Massachusetts-Dartmouth · Jonathan Leonard, Saginaw Valley State University · Gloria Lewis, Wayne State University · John Lilley, University of New Mexico · L. N. Long, Pennsylvania State University · Robert A. Lucas, Lehigh University · Sharon Luck, Pennsylvania State University · Daniel D. Ludwig, Virginia Polytechnic Institute and State University · Arthur B. Maccabe, University of New Mexico · Jack Mahaney, Mercer University · Bill Marcy, Texas Tech University · Diane Martin, George Washington University · Bruce Maylath, University of Memphis · John McDonald, Rensselaer Polytechnic Institute · Olugbenga Mejabi, Wayne State University · Steve Melsheimer, Clemson University · Craig Miller, Purdue University · Howard Miller, Virginia Western Community College · Andrew J. Milne, The Leonhard Center for Innovation and Enhancement of Engineering Education · Pradeep Misra, Wright State University · Charles Morgan, Virginia Military Institute · F. A. Mosillo, University of Illinois · Stanley Napper, Louisiana Technical University · John Nazemetz, Oklahoma State University · James Nelson, Louisiana Technical University · Brad Nickerson, University of New Brunswick · Col. Kip Nygren, West Point Military Academy · Adebisi Oladipupo, Hampton University · Joseph Olivieri, Lawrence Technological University · Joseph Olson, University of South Alabama · Kevin Parfitt, Pennsylvania State University · John E. Parsons, North Carolina State University · Steve Peterson, Lawrence Livermore National Lab · Larry D. Piper, Texas A & M University · Jean Landa Pytel, Pennsylvania State University · Mulchand S. Rathod, Wayne State University · Philip Regalbuto, Trident Technical College · Thomas Regan, University of Maryland-College Park · Larry G. Richards, University of Maryland, College Park · Larry Riddle, Agnes Scott College · Lee Rosenthal, Fairleigh Dickinson University · Isidro Rubi, University of Colorado at Boulder · Vernon Sater, Arizona State University · Dhushy Sathianathan, Pennsylvania State University · Kenneth N. Sawyers, Lehigh University · Carolyn J. C. Schauble, University of Colorado at Boulder · William Schiesser, Lehigh University · Bruce Schimming, Howard University · Keith Schleiffer, Battelle · Murari J. Shah, Technical Graphics · Howard Silver, Fairleigh Dickinson University · Ed Simms, University of Massachusetts · Thomas Skinner, Boston University · Gary Sobczak, Purdue University · Jim Strumpff, Mercer University · Habib Taouk, University of Pittsburgh-Titusville · Massoud Tavakoli, GMI Engineering and Management Institute · Greg Taylor, Northern Arizona University · Ronald Terry, Brigham Young University · Ron Thurgood, Utah State University · Stephen Titcomb, University of Vermont · Joseph Tront, Virginia Polytechnic Institute and State University · Israel Ureli, Ohio University · Bert Van Grondelle, Hudson Valley Community College · Lambert VanPoolen, Calvin College · Akula Venkatram, University of California at Riverside · Thomas Walker, Virginia Polytechnic Institute and State University · Richard Wilkins, University of Delaware · Frazier Williams, University of Nebraska · Billy Wood, University of Texas-Austin · Nigel Wright, University of Nottingham

10 Great Engineering Achievements

In celebration of its silver anniversary, the U.S. National Academy of Engineering identified the Ten Great Engineering Achievements accomplished during the organization's first 25 years. Initially selected because they represent major breakthroughs, these achievements have initiated whole new areas of engineering. In the pages that follow, we note the types of design problems faced by interdisciplinary teams of engineers who work in these fields, and we represent contemporary examples in the photographs. Many of the applications you will encounter throughout *The Engineer's Toolkit* are based on these achievements.

1

SPACE TRAVEL

Although 1972 marked the last in a series of moon landings begun in 1969, the Apollo mission laid a foundation for a whole new generation of space shuttle missions that were dedicated to gathering information about the universe and that continue to test the human ability to travel in space.

Design Problems Several key design problems had to be solved to support the Apollo mission to land humans on the moon. The spacecraft required a new inertial navigation system; the lunar lander ascent engine had to work perfectly because there was no backup engine; the spacesuits had to protect the astronauts in a hostile environment and yet be flexible enough to allow movement; and the Saturn V rocket, which powered all the Apollo flights, had to be 15 times more powerful than the biggest rockets available in the early 1960s.

Application Areas Today, rockets launch deep space probes such as the Voyager, which continues to send back images from as far away as Venus and beyond, while space shuttles launch information-gathering satellites and scientific equipment. The Hubble Space Telescope (HST) was launched from the

The space shuttle Endeavour floats above the earth at an altitude of 381 miles, with the west coast of Australia forming the backdrop for the 35mm frame. While perched on top of a foot restraint on the Endeavour's Remote Manipulator System arm, astronauts F. Story Musgrove (top) and Jeffrey A Hoffman wrap up the last of five space walks. They have succeeded in their mission to repair the Hubble Space Telescope.

space shuttle Discovery on April 25, 1990. NASA had designed HST to allow scientists to view the universe up to 10 times more sharply than they could with earthbound telescopes. Unfortunately, scientists soon discovered that the primary mirror in HST was flawed and could not focus properly. It was another space shuttle, the Endeavour, that operated as a kind of Hubble repair station.

Engineering Disciplines Although the fully operational Hubble Space Telescope now stands as an impressive achievement for NASA, the repair job itself was perhaps an equally important achievement. Materials engineers helped develop metals, plastics, and other materials that could withstand the rigors of the launch and the environment in space. Mechanical engineers helped develop the mechanical structures that position HSTs mirrors and other moving apparatus and electrical engineers helped develop the complex computer, communication, and power systems.

2

APPLICATION SATELLITES

Application satellites and other spacecraft orbit the earth to capture, relay, and transmit specific types of information, or to perform manufacturing processes that rely on special properties of the extraterrestrial environment, such as zero gravity.

Application Areas Satellite systems provide information on weather systems, relay communication signals around the globe, survey the earth and outer space to map uncharted terrain and provide navigational information for vehicles on land, in the oceans, and in the air. The Endeavour has been involved in NASA's Mission to Planet Earth, which is designed to help the international scientific community better understand which environmental changes are caused by nature and which are induced by human activity. Throughout 1994 the shuttle orbited the earth with the Spaceborne Imaging Radar-C and X-Band Synthetic Aperture Radar system, which illuminates the earth with microwaves, allowing detailed observations at any time,

regardless of weather or sunlight conditions. (See photo below.)

Design Problems A satellite or spacecraft launching system must be designed to generate enough thrust to escape the earth's atmosphere. Once free, it needs to maintain a stable orbit around the earth. In addition, the hull needs to be light and yet strong enough to withstand the stress of the liftoff.

Engineering Disciplines Space-based inventions, such as the Spaceborne Imaging Radar-C (SIR), and satellites in general are a result of the cooperative efforts of aerospace engineers who help develop the systems that put satellites in space, and of chemical, mechanical, and electrical engineers who assist in the development of the radar and imaging systems for applications such as SIR.

3

MICROPROCESSORS

A microprocessor is a tiny computer, smaller than your fingernail, that combines the control, arithmetic, and logic functions of large digital computers.

Application Areas With its small size and powerful capabilities, the

A technician working on the Bit Serial Optical Computer (BSOC), an optical computer that stores and manipulates data and instructions as pulses of light. To enable this, the designers (Harry Jordan and Vincent Heuring at the University of Colorado) developed bit-serial architecture. Each binary digit is represented by a pulse of infrared laser light 4 meters long. The pulses circulate sequentially through a tightly wound 4-kilometer loop of optical fiber some 50,000 times per second. Other laser beams operate lithium niobate optical switches which perform data processing.

applications of microprocessors range from operating remote television controllers or VCR recorders to providing the computational power in hand-held calculators or personal computers. Microprocessors can also be found in communication devices, such as networks that connect computers around the globe, and in automobiles, ships, and airplanes.

Design Problems Key design problems involved in creating microprocessors include miniaturization, increasing speed while controlling the heat produced, and searching for materials stable and reliable enough to store, process, and transmit data at high speeds. For decades engineers have improved the performance of computers by increasing the number of functions contained on the CPU chip. Ultimately this approach created a bottleneck in switching between these functions. In the early 1980s, chip designers addressed this issue and developed a concept of Reduced Instruction Set Computing (RISC) which improves efficiency through the high-

This image of the area around Mount Pinatubo in the Philippines was acquired by the Spaceborne Imaging Radar-C and X-Band Synthetic Aperture Radar system aboard the space shuttle Endeavour in April 1994. This false color image shows the main volcanic crater on Mount Pinatubo produced by the June 1991 eruptions and the steep slopes on the upper flanks of the volcano. The red color shows the rougher ash deposited during the eruption. The dark drainages are the smooth mudflows that continue to flood the river valleys after heavy rains. This radar image helps identify the areas flooded by mudflows, which are difficult to distinguish visually, and assess the rate at which the erosion and deposition continue.

speed composition of an optimized minimal set of instructions. Research and development is on-going in another area of microprocessor design as well: optical computing. A dream since the 1940s, optical computing represents a fundamental change in how switching occurs—through optical signals rather than electronic signals. Since a computer is nothing more than a complex system of on/off switches, the speed at which the switches can turn off and on is the single most critical factor in determining the computer's performance. Engineers can design optical switches that operate well into giga-hertz range, while electronic switches are currently restricted to about 100 megahertz.

Engineering Disciplines An application such as optical computing is a result of many years of collaboration between electrical engineers who

Inspection of the largest, most powerful energy-efficient jet engine, which was designed for the Boeing 777 jets.

design the complex laser systems, computer engineers who work on the computational structures, and chemical engineers who develop the actual lasers.

4
JUMBO JET

Much of the success of jumbo jets (747, DC-10, L-1011) can be attributed to high-bypass fanjet engines, which allow the planes to fly farther using less fuel and with less noise than previous jet engines. Jumbo jets also have an increased emphasis on safety. For example, a 747 has four main landing-gear legs instead of two; a middle spar was added to the wings in the event one is damaged; and redundant hydraulic systems operate the critical system of elevators, stabilizers, and flaps that control the motion of the plane.

The newest jumbo jet is Boeing's 777, which was entirely developed using computer-aided design systems.

Application Areas The technological advances of superior fuels, engines, and materials achieved during the creation of the jumbo jet benefited the space program and the military.

Design Problems A major design problem faced by creators of the jumbo jet was creating an engine and fuel supply that could generate sufficient horsepower and thrust needed to lift the huge aircraft off the ground. The hull had to be strong enough to withstand the stress of the flight, without being so heavy that fuel efficiency was lost. The internal environment had to be comfortable for the passengers, especially during liftoff and landing.

Engineering Disciplines Engineers have been developing increasingly powerful jet engines ever since Englishman Sir Frank Whittle developed the first jet engine prototypes in 1937. The evolution of Whittle's engine into the Boeing 777 engine was largely due to the efforts of mechanical engineers who specialized in dynamics, thermodynamics, combustion systems, and materials.

5
MEDICAL IMAGING

A CAT (computerized axial tomography) scanner is a machine that generates three-dimensional images or slices of an object using x-rays. A series of x-rays is generated from many angles, encircling the object or patient. Each x-ray measures a density at its angle, and by combining these density measurements using sophisticated computer algorithms, an image can be recon-

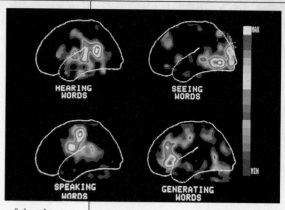

The PET scans shown here reveal localized brain activity under four different conditions, all related to language. Physicians use the PET scanner to diagnose brain and heart disorders and certain types of cancer.

structed that gives clear, detailed pictures of the inside of the object. A PET (positron-emission tomography) scanner reveals locations of intense chemical activity within the body. Sugar labeled with radioactive isotopes that emit particles called positrons is first injected into a person's bloodstream. These positrons collide with electrons made available by chemical reactions in the body. The scanner detects the energy released by these collisions and maps metabolic "hot spots"—regions of an organ that are most chemically active at the time.

Application Areas Doctors use diagnostic radiology to both detect and diagnose diseases, and in combination with other medical techniques, to treat diseases. These devices can also be used to explore the structure of both organic and inorganic materials.

Design Problems The key design problem lies in getting a clear picture without harming the patient or technician. In addition, the machine needs to withstand the various fields it produces.

Engineering Disciplines The PET and similar medical scanning systems are a product of biomedical engineering, a very specialized field of engineering that combines electrical engineering with medicine. Also involved are materials engineers—mechanical engineers who develop the special structures that contain, direct, and withstand the sometimes hazardous electromagnetic and radioactive emissions often used in such systems.

6

ADVANCED COMPOSITE MATERIALS

A composite consists of a matrix of one material that has been reinforced by the fibers or particles of another material. The choice of the composite is determined by the need of the application. For example, does the application require a material that is strong, flexible, stiff, lightweight, heavy, heat-tolerant, porous, dense, or wear-resistant?

Application Areas Advanced composites can be found in most products, including automobile components, communication systems, building materials, artificial joints and organs, and machine parts.

Design Problems Composite material designers must determine how the various materials will interact with each other and how the composites will act over time given the often high-stress situations in which they are used.

Engineering Disciplines Mechanical engineers are most frequently involved in the development of composite materials. Working with chemists and sometimes chemical engineers, mechanical engineers design applications, such as artificial limbs, for particular composite materials.

7

CAD/CAM/CAE

CAD (computer-aided design) refers to computer systems used by engineers to model their designs. These systems may include plotters, computer graphics displays, and 2-D and 3-D modelers. CAM (computer-aided manufacturing) systems are used to control the machinery or industrial robots used in manufacturing the parts, assembling components, and moving them to the desired locations. CAE (computer-aided engineering) systems support conceptual design by synthesizing alternative prototypes using rule-based systems; they also support design verification through rule checking of CAD models. CAD/CAM/CAE systems are intended to increase productivity by optimizing design and production steps, and by increasing flexibility and efficiency.

Application Areas CAD/CAM/CAE systems are used in all engineering design disciplines to support product design, test, and production.

Design Problems Engineers must correctly convert real-world specifications to valid computer models, design appropriate computer tests for the model, and design computer systems to accurately and economically manufacture the final product.

Engineering Disciplines CAM has had a major effect on the work of many industrial engineers. Switching over from labor-intensive manufacturing to computer-aided manufacturing has changed and will continue to change the national and international industrial environ-

Zina Bethune, ballet teacher, has special long artificial hip implants coated with cobalt and chrome. She needed the implants because she suffers from degenerative arthritis.

This automated integrated circuit insertion device, designed by Lockheed, is installing an integrated circuit into a circuit board for a satellite program.

ment. Because they use CAM systems themselves, industrial engineers are heavily involved in their development. Computer and electrical engineers design the computer and control systems and write the software that drive CAM systems.

8

LASERS

Light waves from a laser (that is, light amplification of electromagnetic waves by stimulated emission of radiation) have the same frequency and thus create a beam with one characteristic color. More importantly, the light waves travel in phase, forming a narrow beam that can easily be directed and focused.

Application Areas CO_2 lasers can be used to drill holes in materials such as ceramics, composite materials, and rubber. Medical uses of lasers include repairing detached retinas, sealing leaky blood vessels, vaporizing brain tumors, removing warts and cysts, and performing delicate inner-ear surgery. Lasers are used in scanning devices to scan Universal Product Codes. High-power lasers also are used in weapons and to create 3-D holograms.

Design Problems A key concern is controlling the heat and power generated by lasers so the laser does not harm the patient or the product.

Engineering Disciplines Although they were not heavily involved in the original design and development of lasers in the 1950s, engineers have developed applications for them in

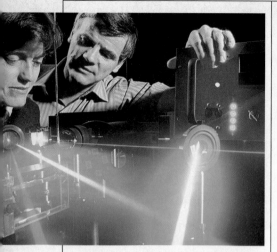

Automatic analysis of DNA is performed with laser beams by Leroy Hood and Jane Sanders, biologists at California Institute of Technology. This DNA sequenator is also called the "Gene Machine."

many areas: Electrical engineers use them in optical fiber communications, civil engineers use them to perform accurate surveying, and mechanical engineers use them for precise cutting of metal parts.

9

FIBER OPTIC COMMUNICATIONS

An optical fiber is a transparent thread of glass or other optically transparent material. It can carry more information than either radio waves or electrical waves in copper telephone wires. In addition, fiber optic communication signals do not produce electromagnetic waves that cause cross-talk noise on communication lines. The first transoceanic fiber optic cable was laid in 1988 across the Atlantic. It contains four fibers that, together, can handle up to 40,000 calls at one time.

Application Areas An increasing number of communications and computer systems are converting to fiber optics due to its enormous information capacity, small size, light weight, and freedom from interference.

Design Problems A key design problem is to create cables that can withstand the stress of being buried under ground or under the ocean. Another design problem is the need to

keep the various signals independent and free from outside interference.

Engineering Disciplines Mechanical engineers design the manufacturing systems that produce glass and plastic optical fibers. Electrical engineers design the transmitters, amplifiers, and receivers that, along with the fibers, carry the optical signals from source to destination.

10

GENETIC ENGINEERING

A genetically engineered product is created by splicing a gene that produces a valuable substance from one organism and placing it into another organism that will multiply itself and the foreign gene along with it. Genes are artificially recombined in a test tube, inserted into a virus or bacteria, and then inserted into a host organism in which they can multiply. Once the new organism has been created, a system has to be designed to produce and process the product in large quantities at a reasonable cost.

The first commercial product of genetic engineering was human insulin, which appeared commercially under the trade name Humulin. The molecules are produced by the genetically engineered bacteria and are then crystallized into human insulin.

Application Areas In addition to creating new drugs and vaccines, genetic engineering has been used to create bacteria that can clean up oil

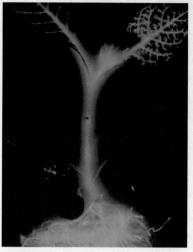

Researchers incorporated firefly gene codes for the enzyme that catalyzes the chemical reaction to release energy in the form of light, into the DNA of a tobacco plant.

spills and detoxify wastes. The process is also used to create genetically altered plants that are pest and disease resistant or have certain desirable characteristics such as improved taste, shipping hardiness, or longer shelf life.

Design Problems Engineers must translate the laboratory work of biologists into the large-scale commercial manufacturing systems that are robust, safe, and cost effective.

Engineering Disciplines Mechanical engineers design equipment for growing large quantities of genetically engineered organisms, chemical engineers design processes for separating out the desired end product, and environmental engineers evaluate the potential impact on the environment.

Thinner than a human hair, an optical fiber can carry more information than conventional radio waves or electrical waves.

Engineering Design and Problem Solving

Steven K. Howell
Mechanical Engineering Department
Northern Arizona University
Flagstaff, Arizona

The Benjamin/Cummings Publishing Company, Inc.

Redwood City, California · Menlo Park, California
Reading, Massachusetts · New York · Don Mills, Ontario
Wokingham, UK · Amsterdam · Bonn · Singapore
Tokyo · Madrid · San Juan

This is a module in The Engineer's Toolkit™, a Ben-
jamin/Cummings SELECT™ edition. The Engineer's
Toolkit and SELECT are trademarks of the Benjamin/
Cummings Publishing Company, Inc. Contact your
sales representative for more information.

Credits:
Chapter 1: pg 1, © Peter Menzel.
           pg 5, Illustration by Orville Wright, *How
           We Invented the Airplane* (New York,
           1988).
           pg 4, © Telegraph Colour Library/FPG
           International.
           pg 6, Courtesy of International Business
           Machines Corporation.
Chapter 2: pg 11, Courtesy of Ford Motor Company.
           pg 29, Courtesy of Autodesk, Inc.
Chapter 3: pg 37, Courtesy of NASA-Ames Dryden.
           pg 49, U. S. Department of the Interior,
           National Park Service, Edison Historical
           Site.
           pg 61, ©ASEE, Journal of Engineering Edu-
           cation, Vol. 83, No. 3, p. 268.
Chapter 4: pg 75, Microscience International Corpo-
           ration.

ISBN 0-8053-6349-1

The Benjamin/Cummings Publishing Company, Inc.
390 Bridge Parkway
Redwood City, CA 94065

Contents

1 An Introduction to Engineering Problem Solving

Communications Satellites

Engineers' design the products, technologies, devices, and systems that define and influence the way we live. Push the right buttons on your telephone and you can speak as clearly with someone on the other side of the world as with your next door neighbor. The system that provides this service is a network of earth-orbiting communications satellites that connect billions of homes and businesses around the world. To successfully design these satellites, engineers had to solve significant problems such as reducing the size and weight of the electronic circuitry, creating a reliable system to get power to the satellite, and protecting the delicate electronics from extremes of heat and cold. Since solving problems is central to engineering, this module presents a systematic method for solving technical problems.

INTRODUCTION

Technology permeates nearly every aspect of our lives. High-speed jet travel, modern telecommunications, computer-aided engineering and graphics, and routine space flights were subjects of science fiction just 50 years ago. Yet all these accomplishments that we take for granted were created, designed, and produced by engineers. Engineers apply technology to meet a human need. Engineers solve problems. Since problem solving is fundamental to the study and practice of engineering, you need to develop a systematic and organized approach to problem solving early in your program of study. This module presents a general approach to solving technical problems that you can use for most problems you will encounter as an engineer.

1-1 WHAT IS ENGINEERING?

Since you are reading this book, you probably are interested in a career in engineering or are at least considering the study of engineering. At this point in your engineering career you probably have little experience in the practice of engineering and may not even know what engineers do. You may have selected this field because of advice from an engineering acquaintance, counselors, parents, or friends. When considering an engineering career, you should begin by learning what engineering is and what engineers do.

As an engineering student, you will be taking a course of study strong in mathematics and science. While your studies may be similar to those of a chemistry, physics, or mathematics student, your activities as an engineer will be quite a bit different than those of a scientist.

Engineers apply principles of math and science to solve technical problems. The laws and forces of nature are directed by engineers to meet human needs. For example, the combustion of hydrocarbon fuel (gasoline) releases heat (a form of energy). This chemical process has been harnessed by engineers to power your automobile. The heat energy released by combustion moves a piston in a cylinder, which in turn rotates the crankshaft of the engine. Countless technical problems had to be solved by various engineers before this engine became a working reality. Engineering is essentially a creative profession, since engineers are creating or synthesizing new devices, products, or systems.

Therefore, we can say that engineering is the application of science and mathematics to solve technical problems and create new systems, products, or devices to benefit civilization. The Accreditation Board for Engineering and Technology (ABET) gives the following definition of engineering:

Engineering is the profession in which a knowledge of the mathematical and natural sciences gained by study, experience, and practice is applied with judgment to develop ways to utilize, economically, the materials and forces of nature for the benefit of mankind.

1-2 THE TECHNOLOGICAL TEAM

Today's technological problems are too complex to be solved by one individual engineer. Creating a complex machine such as the internal combustion engine requires a specialized understanding of nearly all engineering and scientific disciplines. Technological problems are solved by a team of people, each member contributing specialized knowledge from their expertise. As an engineer you will be solving problems as part of a technological team (see Figure 1-1). The technological team consists of people from many scientific fields in the following categories:

- Scientists
- Engineers
- Technologists
- Craftspeople

Each member of the technological team has training and abilities that span the technological spectrum (see Figure 1-1). Scientists' training is at the theoretical end of the spectrum. At the other end of the spectrum, the implementation level, are the artisans or craftspeople. This group consists of people skilled in the trades and responsible for the actual construction, fabrication, and implementation of an idea. The success of any engineering project depends on the cooperation and performance of the entire team. As part of a technological team, you need to understand and appreciate the training and skill of each team member. We will now look at the role of each team member in more detail.

Scientists

While scientists and engineers have similar foundations of study in mathematics and science, they differ in the questions they ask and the net results of their studies. *Scientists* ask the "why?" questions, while engineers focus on the "what can I make with it?" questions. Scientists want to understand why our world behaves the way it does, while engineers primarily apply scientific principles to solve problems or meet human needs. Like engineers, scientists have strong training in mathematics and science, but scientific study emphasizes the theoretical. Scientists typically have strong laboratory backgrounds, conducting experiments to discover new knowledge (see Figure 1-2). In contrast, engineers apply established scientific theories and principles to create new products or solve technical problems. Although scientists are usually involved with obtaining knowledge about the natural world, their roles sometimes overlaps those of engineers.

Figure 1-1
The technological team and technological spectrum

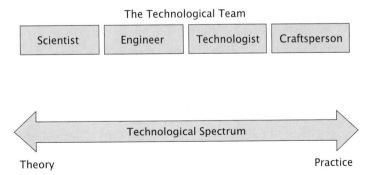

Figure 1-2
Scientists conducting research to discover new knowledge

What roles would scientists play in the development of an internal combustion engine? They would develop an understanding of the combustion process defining the heat energy released when the fuel (gasoline) is burned with air. Scientists might also evaluate methods to reduce the combustion emission products (air pollution). They could have a part in developing new materials for engine parts or they might play a role in determining physical properties of lubricants and thereby reducing wear in the engine.

Engineers

Engineering is called an *applied science* because it attempts to do something useful with scientific theories and principles. *Engineers* are the link between scientific theory and the implementation of technology. The end result of science is *new knowledge.* The end result of engineering is *design.* Design is a creative process that results in a new device, system, structure, or process that satisfies a specific human need. Design can also be called *invention.*

Although engineers are trained in scientific theories, they may create or build a device that works without fully understanding why it works. The Wright brothers, for example, built and flew the first airplane based on experience they gained building bicycles, not from a knowledge of the science of aerodynamics. They knew that an airplane could not be too stable or the pilot would not be able to control it. This knowledge was not based on scientific research, but on their experience with the sideways instability of the bicycle—a device the Wrights were intimately familiar with. The Wright brothers solved the problem of control by devising an ingenious system of wing warping using cables that twisted the wing (see Figure 1-3). This system allowed their first aircraft to fly straight and make banked turns. Their ideas eventually led to the use of ailerons on modern aircraft. It wasn't until 30 years later that aerodynamicists (scientists) developed the theory necessary to calculate conditions of optimum stability for an airplane.

Figure 1-3
Wing warping in the first Wright airplane

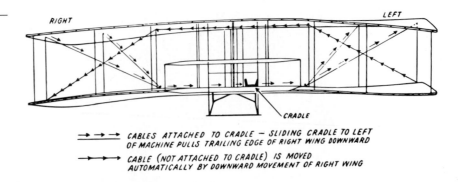

Engineers play a key role in devising the mechanisms that convert the heat energy released when gasoline is burned in an internal combustion engine into mechanical work. Engineers determine the proper sizes and geometrical relationships of all the components making up the engine. They create the systems that remove excess heat from the engine.

Technologists

Engineers and technologists have overlapping functions in the practical or implementation level. *Technologists* have basic training in science, mathematics, and engineering, but not at the same depth or level as an engineer or scientist. Technologists generally do not have the theoretical background of an engineer. They receive more training in manual skills pertinent to the construction and implementation of projects. A technologist may have either a baccalaureate degree or a two-year associate degree.

Technologists would do much of the actual implementation of the ideas of engineers for the design of an internal combustion engine. They would be responsible for most of the drafting or producing the engineering drawings. Any testing or troubleshooting, including data reduction and recording, would be done by technologists. Technologists might also direct the machinists and craftspeople who actually fabricate the parts of the engine (see Figure 1-4).

Craftspeople

Craftspeople are at the practical or implementation end of the spectrum and primarily manufacture or assemble the products designed by the rest of the team. They may not have much training in the principles of science and engineering, but they possess the skills required to operate machinery and to fabricate and construct parts. Craftspeople are usually trained on the job through some sort of apprenticeship. They have skills in the industrial arts, such as welding, electrical wiring, construction, plumbing, and operating special machinery.

Craftspeople would create the molds and dies required to mold the block of an internal combustion engine. They also would operate the lathes and mills used to make the valves and other components of the engine. Any welding and wiring electrical wiring would be done by appropriate craftspeople.

Figure 1-4
**Technologist
assembling electronic
equipment**

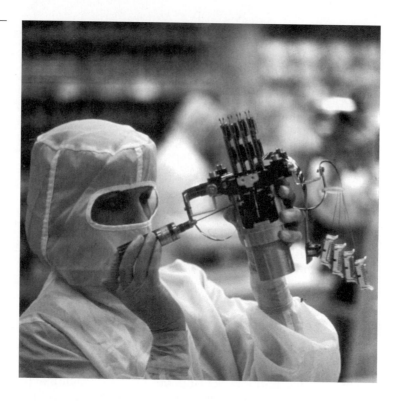

1-3 ENGINEERING PROBLEM SOLVING

Problem solving is the foundation of all engineering activities. As an engineering student, you will spend a significant amount of your time solving problems. You will start with a foundation in mathematics and science, where the problems are clearly defined and you must specify what are the known and unknown quantities. These problems usually include the proper equations, parameters, and assumptions. As you move into more advanced engineering courses, you will solve open-ended design problems, which may be vaguely defined and have many correct solutions.

Since skill at problem solving is fundamental for engineering, it is essential that you develop a systematic methodology for problem solving. To determine a path or method that fits the specific problem you're trying to solve, it may be useful to categorize engineering problems into two broad categories: closed-ended or *analysis problems* and open-ended or *design problems*.

Before making a distinction between the two classes of problems, we need to define the *properties* of a system. The properties are the measurable or observable characteristics of the system such as size, weight or mass, temperature, velocity, and so on. When you calculate or measure the properties of a system, you are solving an analysis problem. Generally analysis problems have a single correct solution, while design problems have multiple correct solutions. For example, we can calculate the weight (a physical property) of a hammer given its physical dimensions and density of material. There is only one solution to this problem for a given hammer.

Design problems, however, have many correct answers. Design is the creation of a device or system that has given properties, while analysis is the determination of the properties of a given device or system. A design problem would be to create a hammer with a given weight. There are an infinite number of hammers that have a specified weight. Engineering design is a process, or sequential methodology, that has the result of producing a device, structure, or system that satisfies a need.

Design is the activity that distinguishes engineering from other technical fields. Engineers have technical knowledge and, using creativity, apply that knowledge to solve a problem or meet a need. Design is an important aspect of all fields of engineering. Mechanical engineers design and fabricate new machines, civil engineers build new bridges and structures, and electrical engineers design new microchips and electronic systems.

Often the difference between the two classes of problems is in the problem definition statement, as illustrated by the following example.

EXAMPLE 1-1　　## An Analysis Problem

To illustrate single-answer analysis problems, consider this problem statement from a common textbook in statics:

> A 600-pound crate is supported by the rope-and-pulley arrangement shown in Figure 1-5. Determine the force in the rope.

SOLUTION

In this example the system or device is specified, and we are asked to determine a property of the system (the tension in the rope). The problem statement is clear and has only a single correct solution. If we know the governing force equations, we can easily understand the problem and solve it by simply substituting appropriate values into the equations. The solution to this problem is the value for one parameter—the force in the rope. The device used to lift the crate is defined in the problem statement as a rope and pulley assembly.

. . . .

EXAMPLE 1-2　　## A Design Problem

The previous problem of calculating the rope force can be made into a design problem by slightly modifying the problem statement. By making the problem definition a little more general and asking you to create a device that has specified properties this becomes a design problem:

> Design a device or system that can raise a 600-pound crate to a level of 4 feet with minimum human effort.

Figure 1-5
**Determining the
force in a rope**

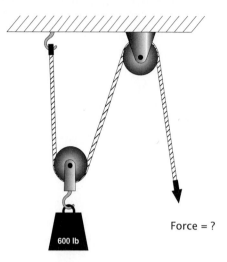

Force = ?

600 lb

SOLUTION

Although this problem statement could produce the same answer as
Example 1-1, it could also be satisfied by a multitude of correct answers.
The opening statement is more general and vague than in the analysis
problem. Some of the properties of the system are specified (the weight of
the crate and the vertical distance), but no device or system is given with
these properties. A correct solution to this problem depends on the information and properties specified in the problem statement. The design
problem is not as clearly defined and lacks some required information.

What additional information would we need to solve this problem? The
size of the crate is an important piece of information and will determine
the type of machine we design. Is cost or economics important? Since
many design solutions must be sold at a profit, the cost of the solution
and market potential will determine the solution. This problem statement
also says nothing about safety. Will it be operated by an individual or a
group of people? As you begin to ask these types of questions, you can
see that the solution to a design problem is very dependent on the information given and the questions asked by the designer. A correct solution
to a design problem requires us to collect information, make assumptions, and fill in missing information. The solution to this problem could
be a simple rope-and-pulley system, or it could be something as sophisticated as a hydraulic lift.

. . . .

1-4 A PROBLEM-SOLVING METHODOLOGY

Since skill at solving problems is fundamental for engineering, it is essential that you develop a systematic methodology. Determining a solution to
simple, clearly defined problems may seem intuitive and not require a formal procedure. However, the types of engineering problems you will
encounter in practice are seldom clearly defined or solved by intuition.
Effective problem solving requires a logical and organized methodology.
While a formal method is no substitute for experience and common sense,

it will allow you to be more effective in reaching a "good" solution to a problem.

In the following chapters you will be introduced to a formal methodology or step-by-step procedure for solving both analysis and design problems. You will use basic steps or procedures to help you organize and arrive at a solution to most problems you will encounter as an engineer. You will begin by looking at analysis problems, which start with a given system and require you to determine an unknown property or properties of the system. In Chapter 3 you will examine design problems, which require you to create or invent a system that possesses specified properties.

The problem-solving methodology you will be using is general in approach and can be adapted to a variety of engineering problems. The actual details of carrying out a solution depend on the nature of the problem being solved. Finding a solution to a textbook physics problem will differ from finding a solution to an open-ended design problem.

SUMMARY

This chapter introduces the types of problems solved by engineers. These problems are usually solved by a team of specialists in technical disciplines. Engineering problems can be divided into two broad categories: design problems and analysis problems. Analysis problems require calculating or measuring the physical properties of a given system. Analysis problems are generally well defined and have a single correct solution. In contrast, design problems are open ended and have many correct solutions. A design problem solution usually requires the creation or invention of a system that has specified properties. The solution to a design problem is therefore a product, machine, or system that meets a human need. Both classes of problems require a methodology or step-by-step procedure to solve. We will present this procedure in detail for both types of problems in later chapters.

Key Words

analysis problem
applied science
craftsperson
design
design problem
engineer

engineering
invention
property
scientist
technologist

Exercises

1. Technology impacts every aspect of our lives today. Consider an airport: a metal detector checks for weapons in the terminal; an elevator transports you to different levels in the building. List ten more inventions that are a part of a modern airport.

2. Research and write a report on recent applications of lasers.

3. Research and write a report on one of the following topics. Identify and discuss some of the analysis and design problems that had to be solved:

    a. the Wright brothers' first airplane

    b. supersonic air travel

    c. the Japanese bullet train

4. Research and write a report on the uses of technology in medical imaging.

5. Interview an engineer and find out what his or her work functions are.

6. Interview a craftsperson (such as a machinist, welder, electrician, or carpenter) and find out what his or her work functions are.

7. Interview a scientist and find out what his or her work functions are.

8. An automotive technician has a detailed knowledge of mechanical systems, electronics, and computer technology. Explain why an automotive technician is not an engineer.

2 Solving Engineering Analysis Problems

Internal Combustion Engine Internal combustion engines are complex machines that convert the energy released from burning gasoline into work to power your car. As the combustion process occurs in the engine, the pressure and temperature increase inside a piston-cylinder arrangement. The increasing pressure causes the piston to move in the cylinder, which in turn rotates a crankshaft in the engine block. Before they can build a working engine, engineers must solve many technical problems to determine pressures, forces, and temperatures inside the engine cylinder. These problems are classified as analysis problems. Engineering analysis is the application of scientific and mathematical principles to determine an unknown quantity or to answer a technological question. In this chapter you will use a five-step problem-solving process to solve engineering analysis problems.

INTRODUCTION

As an engineer, much of your time is spent solving problems. You may reach a solution to simple problems intuitively, without following a formal process. However, as the problems you solve become more complex, you will find that you can reach a solution more efficiently by breaking the problem down into smaller steps. While a formal method is no substitute for experience and common sense, it will allow you to be more effective in reaching a "good" solution to a problem. This chapter introduces a formal step-by-step process you can use to solve engineering analysis problems.

2-1   SOLVING ANALYSIS PROBLEMS

Analysis is the application of mathematical and scientific principles to solve a technical problem. As discussed in Chapter 1, an analysis problem differs from a design problem in that the system is specified and you are asked to determine the properties of the system. In contrast, a design problem (which is discussed in Chapter 3) asks you to determine a system that has specified properties. Generally, the solution to an analysis problem is unique and either right or wrong.

The problem-solving methodology you will use in this text has been broken down into five basic steps:

1. Define the problem.
2. Collect information.
3. Generate a solution.
4. Refine and implement a solution.
5. Verify and test the solution.

To understand these five steps, try the following simple analysis problem taken from a standard physics textbook. You will apply principles of physics to calculate the maximum height of a vertically tossed snowball projectile.

Application 1:   **DETERMINING THE MAXIMUM HEIGHT OF A PROJECTILE**

Given the initial mass, velocity, and height of a snowball, you will apply a five-step procedure to determine the maximum vertical height reached by the snowball.

 1. Define the Problem

The first step is to state the problem clearly and unambiguously. Since this is an analysis problem, there should be one correct answer to this problem: the maximum height traveled by the snowball. The problem statement, therefore, is

A snowball is tossed straight up in the air with an initial velocity of 100 ft/sec. The moment it leaves the thrower's hand it is 6 feet

Figure 2-1
Sketch of the snowball problem

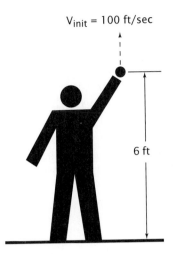

V_{init} = 100 ft/sec

6 ft

above the ground. What is the maximum height reached by this snowball before it returns to the ground?

The problem statement must clearly tell what is unknown and what is known. For simple problems such as this one, a statement of what is known, or given, followed by a statement of what quantities need to be determined are sufficient.

The problem statement should also include a diagram or sketch, as shown in Figure 2-1. A sketch will help you organize your thoughts and help you communicate the solution process to someone else. The type of sketch you use depends on the nature of the problem. For example, to solve an electrical circuit problem you would use a schematic diagram and label all known values of resistors, inductors, capacitors, voltages, and currents. For a mechanics problem you would use a free-body diagram, which shows values, such as forces, magnitudes, directions, and masses.

2. Collect Information

Before you continue in the solution process, you need to collect and substantiate all physical data pertinent to the problem. These quantities may include dimensions, voltages, currents, temperatures, masses, velocities, and displacements. We will discuss sources of information in more detail later in this chapter.

Besides collecting physical data pertinent to the problem, you need to develop an understanding of the theory or principles relevant to the problem. Understanding the relevant theory is vital to determining a proper solution path for your problem. You also need to identify limitations of the selected theory. For example, to solve the snowball problem, you will use a theory that ignores the drag on the snowball due to air friction. As you solve a variety of engineering problems, you will learn which theory applies to your problem and the limitations of that theory. Having a good understanding of the theory will help you identify the important physical data needed to solve the problem.

To solve real problems, usually you make *assumptions* because you don't have access to all the physical data. As you gain experience solving engineering problems, you will develop the understanding necessary to make valid assumptions that will not significantly affect the accuracy of the solution. For example, mechanics problems usually assume that the acceleration of gravity is a constant, but in reality it varies according to the elevation and location on the earth. The solution accuracy of most mechanics problems is not greatly affected by making this assumption.

The solution to the snowball problem applies Newton's laws of motion, which means you need to have a good foundation in physics. Other information you might need to solve this problem includes:

- Acceleration of gravity—or do you assume a constant value?
- Mass of the snowball.
- Size of the snowball.
- Aerodynamic drag coefficient.
- Prevailing winds—or do you ignore them?
- Initial velocity of the snowball.

Using a textbook in physics, you could derive an equation of motion for this snowball from Newton's laws as

$$h(t) = -16t^2 + 100t + 6$$

This model gives the height h as a function of time t. The -16 comes from the acceleration of gravity, which is only approximate and is assumed to be a constant. The $100t$ term is derived from the initial velocity, and the 6 comes from the initial height. This formula is not an exact description of the snowball motion but is an *approximate model*. It assumes no air resistance or wind and a constant acceleration due to gravity. If air resistance, wind, or a variable acceleration from gravity is added to the description of this model, the model will become more accurate but also more complex. As you gain experience in problem solving, you will develop an understanding of what level of solution accuracy is required. As a rule, you should start with a simple model and progress to more complexity if you need a greater level of solution accuracy.

Later in the problem-solving process, you will verify and test this solution. You may discover that a simplified model does not give the level of solution accuracy you need. At that point the solution process becomes iterative: You will go back to step 2 and repeat the solution process with a more complex model. You may include the terms ignored for the first pass: the air resistance and a nonconstant acceleration of gravity.

3. Generate a Solution

Having collected all the information required for this problem, you need to evaluate and then assimilate this information into the solution process. For a simple problem such as the snowball trajectory, step 3 may just involve substituting values into equations. A more complex problem

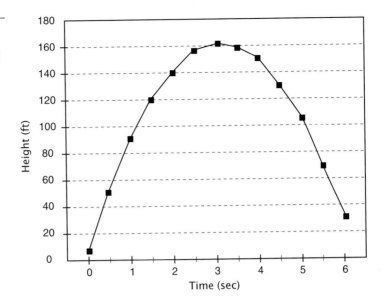

Figure 2-2
**Graphical solution:
height of the snowball
as a function of time**

might require writing a computer program, generating plots, or conducting an experiment to derive the answer.

For the snowball problem, you will start by using the equation given in step 2 that describes height as a function of time. This solution ignores air resistance and assumes the acceleration of gravity is a constant. Later in the analysis process, as you verify and test the solution, you may find it necessary to adopt a more complex model.

If you adopt the simplified model for snowball height, there are several possible techniques or paths you can use to determine the maximum height. One way to solve a problem such as this is to use a *graphical solution*, plotting the height *h* as a function of time *t*. For a problem in which one quantity depends on another quantity (most engineering problems are like this), generating a plot is often the easiest way to solve the problem. The graph in Figure 2-2 shows that the maximum height is approximately 160 feet.

Alternatively, you can calculate the maximum height by applying principles of physics. A fundamental law of physics is called the *conservation of energy*. This law states that the energy possessed by the snowball at the beginning of its trajectory (when the velocity is 100 ft/sec) must be equal to the energy possessed at the maximum height (when the velocity is 0). There are two types of energy possessed by the snowball, *kinetic energy* and *potential energy*. You can think of kinetic energy as motion energy, or the energy possessed by a body of mass *m* moving with some velocity *v*. The formula for kinetic energy can be found in any physics book:

$$KE = \frac{1}{2}mv^2$$

The potential energy is the gravitational energy possessed by a mass *m* raised to some height *h* against the force of gravity. The formula for potential energy from a standard physics book is

$$PE = mg(h - h_0)$$

where g is the acceleration due to gravity and is equal to 32.2 ft/sec², and h_0 is the initial height of the snowball (6 feet).

Using the principle of conservation of energy, you can say that the potential energy possessed by the snowball must always be equal to the kinetic energy. Since you want to solve for the maximum height h, you can equate PE to KE and substitute 100 ft/sec² for the value of v. To apply conservation of energy, you equate the potential and kinetic energies:

$$KE_{max} = PE_{max} = \frac{1}{2}mv^2 = mg\,(h - h_0)$$

You can cancel out m because it is on both sides of the equal sign and apply algebraic relationships to solve for h. Rearranging and substituting values for the variables you have

$$h = \frac{v^2}{2\,(g)} + h_0 = \frac{100^2}{2\,(32.2)} + 6 = 162.2$$

This answer is calculated directly from the conservation of energy and basic physical relationships for kinetic and potential energy. Figure 2-2 shows an answer of approximately 160 feet for a maximum height, which occurs at about 3 seconds. The graphical solution is in agreement with the answer obtained by the principles of physics.

As you can see by this simple illustration, there are at least two paths or techniques you can use to derive an answer. Each technique offers certain advantages. The graphical solution is simple and doesn't require an understanding of physics or calculus. However, it yields an approximate answer. The answer obtained by conservation of energy is more precise (within the limits of the assumptions) but requires an understanding of physics.

Many complex problems are too difficult to solve with an exact mathematical expression, and you need to use graphical or other approximations to obtain a solution. The accuracy of the graphical solution depends on the size and detail of the graphs.

4. Refine and Implement a Solution

Step 4 is the phase in which you actually derive an answer to the model, either by substituting values into the relevant equations or by plotting a graphical solution. Simple problems such as this one may not have separate steps 3 and 4: The solution implementation simply involves substituting values into appropriate equations. This step was completed in step 3 for the snowball problem. Substituting values into the relevant equations or reading the solution from the graph of Figure 2-2 is the solution implementation.

If you have to write a computer program to solve your problem, step 4 might be the actual coding and debugging of the program. For the snowball problem, refining a solution might consist of recalculating the trajectory taking wind and air resistance into consideration.

5. Verify and Test the Solution

An inexperienced problem solver may be tempted to stop after step 4. However, just because you obtained a solution does not mean that solution is accurate or even realistic. The solution is not complete until it has been checked for mathematical accuracy.

If possible, you should verify the solution. This verification may involve designing an experiment and collecting data to compare with the analytical solution. Sometimes you can rework the problem using a different theory or technique and then compare the answers. For example, you got the same answer to the snowball problem with both the graphical technique and the conservation of energy method. You can also estimate the answer to verify your solution. Estimation is discussed in detail later in this application.

You also need to check all your calculations for *dimensional consistency*. You should substitute all the dimensional units associated with the numerical values, such as feet, seconds, pounds, and so on, into the proper equations, and you should check to see if the answer is in the appropriate units. This process is illustrated later in this application.

Units and Dimensional Consistency

Usually the numbers used in solving engineering problems have dimensions associated with them. Dimensions of length, mass, temperature, or other standards are used to characterize these numbers. When you specify a physical quantity such as length or temperature, it is not sufficient to simply give a number. The magnitude defining a physical quantity must be compared with a reference value, which is called a unit. For example, the meter is a fundamental unit referencing length. All physical units or dimensions are classified as either *fundamental* or *derived*. Fundamental units are length, time, mass, force, and temperature. These units are described in Table 2-1.

Table 2-1 Fundamental Units

Length (L)	Length is the fundamental linear dimension between two points and may have units of meters or feet.
Time (t)	Time is the fundamental dimension defining the interval between two events and has units of seconds.
Force (F)	Force is defined through Newton's second law and is the push or pull required to accelerate a mass at a given rate. Force may be directly applied to accelerate an object, such as the force pushing an accelerating automobile. Force may also be applied to an object by fields such as gravity, electrical, or magnetic.
Temperature (T)	Temperature is a dimension relating the hotness or coldness of an object. Temperature units are commonly used to measure the degree of hotness of an object with respect to some reference point. Common temperature scales are the degrees Fahrenheit (°F) scale or degrees Celsius (°C) scale. These scales reference the temperature of an object to the boiling and triple point of water. Other temperature scales are the Rankine (°R) and Kelvin (°K), which are called absolute temperature scales and reference the temperature of an object to absolute zero.

You can derive all other dimensions and units using combinations of the fundamental units. An important part of solution verification is checking for dimensional consistency between all values used in a problem.

After substituting numerical values into the relevant equations, you assign units to those equations. All units should show consistency across both sides of the equation. For example, the equations you derived in Step 3 of the snowball problem describe the height of the snowball:

$$h = \frac{v^2}{2\,(g)} + h_0$$

With the units substituted for v, g, and h_0 in the equation, you have

$$ft = \frac{ft^2/sec^2}{2\,(ft/sec^2)} + ft$$

With units of ft/sec for velocity and ft/sec^2 for the acceleration of gravity, the left and right sides of this equation have dimensional consistency, giving units of ft for the answer.

Estimations and Approximations

Modern computational tools such as pocket calculators and digital computers have given us the ability to display answers to technical problems with several digits of precision. Graphical software allows engineers to create elegant graphs, figures, and reports easily. Unfortunately, the solution to a problem can be visually impressive but still be incorrect. An erroneous result is often accepted and obscured because it is presented with elegance. Before using hand-held calculators, engineers used slide rules as their primary computational device. Engineers knew that their slide rules would only give a solution within three significant figures, so they used *estimation* to verify their answers. Estimation (or approximation) is determining an answer which is "close" to the exact solution. The skill of estimation has largely been lost with the advent of digital computational tools.

An experienced engineer can probably make approximations with an accuracy of 10 or 20 percent of the exact solution. Engineering estimations and approximations can provide you with a basis for judging the validity of an analytical answer. You can discover many computational mistakes and errors in unit conversions by comparing an analytical answer to a simplified estimation or approximation.

You derived an analytical answer to the snowball problem of 162.5 feet for the maximum height. Is this a reasonable answer based on your experience and judgment? The following discussion illustrates approximation based on simple assumptions and common sense:

> The snowball is released vertically, and you know the velocity at release is 100 feet per second. You know that the force of gravity is acting against the snowball's vertical travel. This force will eventually result in a change in direction (velocity equals 0) at the maximum height. You also know that the acceleration due to gravity (downward direction) is 32 feet per second^2. This downward acceleration means that the velocity of the snowball decreases by 32 feet per second for every second of flight. Since the velocity will equal 0 at the maximum height, approximately 3 seconds of time will be required to reduce the initial 100 feet per second velocity to 0 feet per second ($100/32 \cong 3$ seconds).

For a quick approximation of the maximum distance traveled by the snowball in 3 seconds, you will simply multiply the average velocity by the time of travel, 3 seconds. The average velocity is (100+0)/2 = 50 feet per second. 50 feet per second times 3 seconds gives 150 feet traveled. Since this number is close to the value of 162.5 feet per second, you can be reasonably sure that your calculations are correct.

What If

What happens to the maximum height of the snowball if you change the mass of the snowball from one pound to two pounds? Does your answer change? Is the mass an important variable for this problem?

The analysis you performed on the snowball ignored wind and air resistance. If you changed the problem to determine the height of a foam ball (foam is much less dense, so a given-sized foam ball will weigh less than the same-sized snowball) rather than a snowball, would this be a valid assumption? Why not?

As you develop experience at problem solving, you will learn new tools and techniques for solving engineering problems. Your calculus and science courses will introduce you to the principles of mathematics and science that you can apply to engineering problems. The following application illustrates how to break a problem into these five steps.

Here are the five steps to problem solving we will use in this text:

1. Define the problem clearly.
2. Gather relevant information, including the theory, assumptions, and physical quantities.
3. Generate and evaluate a potential solution using the tools and techniques you have available.
4. Refine and implement a solution.
5. Verify and test the solution.

Application 2:   **DISTANCE TRAVELED BY A BICYCLE**

Apply the five-step-problem-solving process to the solution of the following problem.

 1. Define the Problem

A bicyclist begins pedaling a bicycle from a complete stop to a constant speed. She accelerates at a constant rate until she is pedaling at 60 revolutions per minute (cadence, or the number of pedal turns per minute). The biker does not achieve a constant speed until she's been pedaling for 10 minutes. The bike is geared with a 2:1 ratio, meaning that the rear wheel of the bike turns one revolution for every two revolutions of the pedals. How far does the bike rider travel during a 30-minute duration?

Figure 2-3
**Basic geometrical
relationships for a
bicycle**

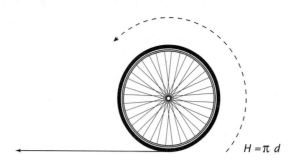

$H = \pi \, d$

2. Collect Information

Collecting information for this problem requires that you understand basic geometrical relationships. The relevant theory for this problem is the relationship between horizontal distance traveled by the bike and the circumferential distance around the wheel. You also need to know the relationship between the circular motion of the pedals and the circular motion of the wheel. A sketch or figure illustrating the basic geometrical relationships between diameter and horizontal distance is shown in Figure 2-3.

Before you can calculate the total distance traveled by the bike, you need to know the rider's speed as a function of time. The problem statement only refers to the change in speed, or the acceleration during the 30-minute interval. Since much of the information needed to solve the problem is not given, you need to make some assumptions.

You also need to know the circumference of the wheel. The problem statement gives you no information about the size of the wheel. Most problems you will encounter as an engineer do not give you every piece of information you need to solve the problem. If you cannot obtain the missing information, then you need to make reasonable assumptions for data that is not available. In this problem it is not unreasonable for you to assume the wheel diameter is 27 inches, the standard size for an adult bike. For every revolution the wheel makes, the bike will travel a horizontal distance equal to the circumference. Therefore, the circumference is $H = \pi d$, where d equals 27 inches. Here is the information you need to solve the problem:

- Diameter d of bicycle tire
- Basic geometrical relationships, circumference $H = \pi d$
- Gear ratio between pedals and wheels
- Rate that biker pedals, revolutions per minute, as a function of riding time
- Relevant conversion factors and units (such as inches, feet, miles per hour, and revolutions per minute)

3. Generate a Solution

The problem statement asks you to determine the total distance traveled during a period of 30 minutes. Because the biker's speed changes during this time interval, you can't simply multiply speed by time to determine distance. This type of problem lends itself to a graphical solution. A plot or graph is invaluable for solving problems in which some quantity depends on some other quantity. The problem statement says the rate of speed change is constant between 0 and 10 minutes. This statement implies that a straight line will connect those points between 0 and 10 minutes. From the problem statement you do know that the speed of the bike doesn't change after 10 minutes. You also know that the initial speed is 0 and increases at a constant rate until 10 minutes. Figure 2-4 shows a graphical representation of the bicycle speed as a function of time.

To solve this problem, you can complete a graph and calculate the area under the curve of speed versus time. Alternatively, you can apply principles of calculus and integrate the function defining speed over the time interval. Since integral calculus is beyond the scope of this text, you will use the graphical solution.

4. Refine and Implement a Solution

Before you can generate a graphical solution, you need to calculate the speed of the bike at different times. Since the information available in the problem statement is in mixed and inconsistent units, you need to verify units and check for dimensional consistency before you can obtain a valid solution. The wheel diameter is given in units of inches (in), the rate of rotation is calculated in revolutions per minute (rpm), time is in minutes (min), and the answer must be in miles per hour (mph).

The information provided in the problem statement does not give the speed of the bike. From the geometry of the problem, you know that the distance traveled by each turn of the wheel is equal to the circumference of the wheel. Therefore, the horizontal distance traveled by the bike for each turn of the wheel is

$$D_{max} = \pi 27 \left(\frac{in}{12} \right) \left(\frac{ft}{in} \right)$$
$$= 7.0685 \, ft$$

You also know that the wheel makes one revolution for every two revolutions of the pedals (from the specified gear ratio). After 10 minutes the biker is pedaling at the rate of 60 revolutions per minute. Therefore, the wheel is turning at 30 revolutions per minute after the biker comes up to her cruising speed (after 10 minutes). The speed traveled after 10 minutes is therefore equal to the distance traveled per turn multiplied by the revolutions per unit of time:

$$S_{10min} = 30 \frac{rev}{min} D_{max} ft$$
$$= \frac{212 \frac{ft}{min} \left(60 \frac{min}{hr} \right)}{5280 \frac{ft}{mi}}$$
$$= 2.4095 \frac{mi}{hr}$$

Figure 2-4
**Graphical solution:
bicycle speed as a
function of time**

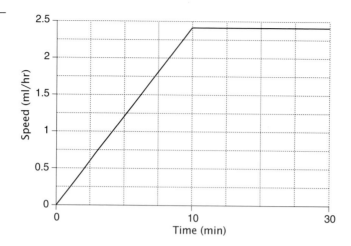

Knowing that this speed is constant from 10 to 30 minutes will allow you to construct the graph shown in Figure 2-4.

The total distance traveled by the biker is the area under this curve. You could find this value by applying principles of calculus and integrating the speed function over the interval from 0 to 30 minutes. However, since you may not have taken a calculus course, you will take the simplified approach and calculate the distance directly from basic geometrical relationships. You will see that basic geometrical relationships can be used directly to calculate the area under this curve. The area under the region from 0 to 10 is simply the area of a triangle, which can then be added to the area of a rectangle from 10 to 30 minutes. The formula for the total distance is therefore the total area under the curve from 0 to 30 minutes:

$$D = \text{Area}_{\text{triangle}} + \text{Area}_{\text{rectangle}}$$

$$= \left[\frac{1}{2}(10-0)\,\text{min}\left(2.4095\,\frac{\text{mi}}{\text{hr}}\right) + (30-10)\,\text{min}\left(2.4095\,\frac{\text{mi}}{\text{hr}}\right) \right] \frac{1}{60} \frac{\text{hr}}{\text{min}}$$

$$= 1.004\,\text{mi}$$

5. Verify and Test the Solution

You applied units and dimensions to the solution in step 4 and derived an answer of 1.004 miles. When you substitute units into the above equations, dimensional consistency is maintained. Is this a realistic and reasonable answer? Applying common sense and approximations, you can verify that this solution is reasonable and realistic. The biker is pedaling at a reasonable rate of one revolution per second. This yielded a speed of approximately 2 miles per hour. Therefore, a total distance traveled of approximately 1 mile during 30 minutes is reasonable and realistic. You can therefore conclude that the solution is accurate.

Try It

Assume that the biker reaches her steady speed after riding 10 minutes, and then she shifts gears from a 2:1 ratio to a 1:1 ratio. Also, assume that the bicycle rider begins to slow her pace after 20 minutes of riding. She is initially pedaling at a constant 60 turns per minute from 10 minutes to 20 minutes of riding time. After 20 minutes of pedaling, she begins a constant decrease in pedaling speed until she comes to a complete stop at 30 minutes. Figure 2-5 summarizes this data. Making the same assumptions as in the original problem, determine the total distance traveled taking into account these new conditions.

Figure 2-5
Bicyclist's cadence as a function of time

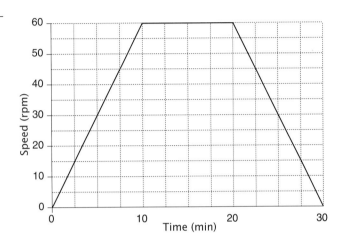

2-2 COMMUNICATING SOLUTIONS

It is important for you to document fully each step in the problem solving process. Engineers never solve problems only for themselves; they must communicate the answers to other people. As a student, you will be solving problems for an instructor, but as a practicing engineer you will be solving problems for a client, a supervisor, a fellow engineer, or the public. Superior analytical skills will be futile if you can't explain the solution to a technical problem to others.

The grade you receive as an engineering student may be strongly influenced by your ability to present solutions to engineering problems in an unambiguous and easy-to-follow format. This chapter presents some guidelines for problem layout and format. A sample layout for engineering problems is shown in Figure 2-6.

These guidelines are intended to provide a suggested format and do not have to be followed precisely. Your instructor may modify or adapt this format to your particular coursework. No matter what layout you use for problem presentation, it must be logical and easy to understand.

1. **Paper and pencil.** Most engineering courses require that problems be written on *engineering paper*. This paper has heading and margin rulings on the front side and is ruled with squares on the reverse side. The squares on the back of the paper show faintly through and provide guidelines for lettering, sketching, and graph construction. Use a medium pencil (2H) so lettering is sharp and not smudged. Erase mistakes completely. Write all lettering in all capital block letters, not in script. Legible lettering is essential.

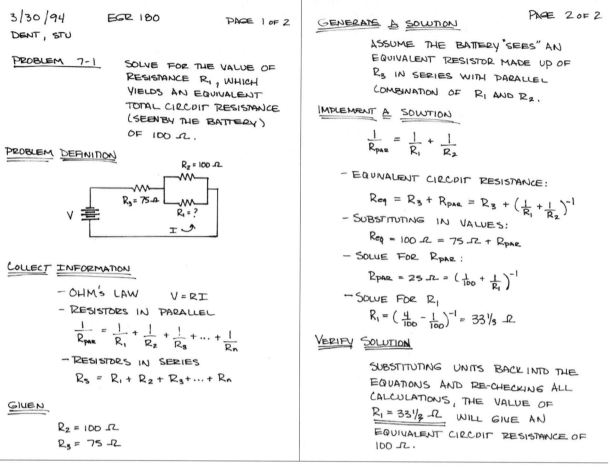

Figure 2-6 Sample problem presentation

2. **Title information.** Center the course title and problem numbers in the title area of the engineering paper. Generally, place the date on the left side of the title and your name on the right side. This is only a suggested guideline; your instructor may have different standards.

3. **Problem statement (including necessary diagrams).** You may simply restate the problem from your textbook, or you may rephrase the problem in your own words. Generally, include some sort of diagram or sketch to communicate relevant information about the problem. A diagram can express efficiently what might be difficult to communicate with words. Be sure to state the pertinent information on your sketch such as voltages, currents, forces, directions, and so on.

4. **Statement of theory (including relevant equations).** List the relevant equations needed to solve the problem. Usually a statement of the appropriate equation with a reference is sufficient. For more complex problems, you may need to show how you derived the equations. For example, in the snowball problem you showed the steps to solving for the unknown height from the conservation of energy relationships. Include all steps performed in the derivation.

5. **Statement of what is given or known.** Indicate which quantities are known or given in the problem. You may include this step with the problem statement.

6. **Statement of assumptions.** Very seldom does the problem statement provide every piece of information needed to solve the problem. You will most likely have to make assumptions or approximations. Don't take for granted that your reader knows which assumptions you made.

7. **Calculations (including all steps).** List every step performed in reaching a solution. In an academic situation the instructor is evaluating your work and needs to see all the steps you performed. If your solution is in error, you will not receive partial credit if the instructor can't follow every step.

8. **Units and dimensions.** A numerical solution without units or dimensions is useless. Part of the verification and testing of your solution is checking for dimensional consistency. Therefore, you must include units and dimensions.

9. **Results and conclusions, including checking for mistakes.** Indicate clearly the final answer. Your solution is not complete until it has been checked for mistakes and reasonableness.

2-3   PROBLEM-SOLVING TOOLS

Skilled carpenters wouldn't think of constructing a house with only a hand saw and hammer. They would use the best tools available to make their work easier and produce a higher-quality finished product. In the same way, your ability to quickly and accurately solve engineering problems will improve as you become skilled in using the various tools available to today's engineers. Much of the training you will receive as an engineering student will be directed to understanding and using these tools. The following section briefly introduces you to some of these tools.

Computer Tools for Engineering Problem Solving

The practice of problem solving has been revolutionized during the past 30 years since the advent of computer technology. As recently as the 1960s, a slide rule, pencil, and paper were the primary tools used to solve engineering problems. The computer revolution has influenced the practice of engineering in much the same way that the invention of the steam engine caused the industrial revolution of the 1800s. Computers have touched nearly every aspect of our lives, including business, transportation, communication, national defense, education, health care, and entertainment. Computing capability has doubled about every two years. Rapid changes in computer technology have placed extra demands on the engineer. As an engineer, you have no choice but to stay abreast of computer developments or risk becoming obsolete. Engineers who are not able to learn how to apply computer technology to engineering problem solving may find it difficult to compete in their profession.

Electronic Spreadsheets   As PC hardware became more powerful, the capabilities of the software also increased. New software tools began to appear in the 1980s that would revolutionize the way engineers solve problems and do design.

One such tool is the *electronic spreadsheet.* The electronic spreadsheet was originally conceived as an electronic ledger for business applications. Other disciplines, such as engineering, have discovered the power of this tool for solving problems. A spreadsheet is an electronic version of the traditional accounting ledger in which the various entries are related to each other. For example, a checkbook is a simple spreadsheet, in which the remaining balance is related to the deposits and debits applied to the account. A spreadsheet consists of *cells,* which are the intersections of row and column entries containing text, numbers, or mathematical relationships. The computational framework of a spreadsheet allows you to define relationships between the various entries and use these relationships to solve mathematical problems. This computational framework is called a *template.* Once you have created a template that defines the relationships between cell entries, you can examine various alternatives by changing the data in the template. You can ask "what if?" questions about your problem. In other words, you could set up a spreadsheet template for calculating the height of the snowball and simply change one value, such as the initial velocity. The spreadsheet model would predict the response of the snowball trajectory to this change. Electronic spreadsheets also give you the ability to present the results of your problem graphically. Sometimes a spreadsheet is the easiest way to generate a graphical plot of your problem results.

Figure 2-7 illustrates a spreadsheet solution to the snowball problem. This figure demonstrates the graphical capabilities of the spreadsheet. Each cell is addressed by its row and column identifier. For example, cell A5 contains the label time(sec) for the column A6 to A20. The column B5 to B20 contains the formula for the height, using the value in the corresponding A column as input. This spreadsheet represents the template for this problem. Once the mathematical relationships have been defined for all elements of the template, it is very easy to ask "what if?" questions. For example, the formula for height in cell A3 could easily be changed to represent a different initial velocity. Changing the velocity will produce an interactive change in all calculated values for the height and then will generate a new graph.

Figure 2-7
Spreadsheet solution to the snowball problem

A	A	B	C	D	E	F	G	H	I
2									
3	Height of Snowball= −16*time^2 +100*time +6								
4									
5	time(sec)	height(ft)							
6	0	6							
7	0.5	52							
8	1	90							
9	1.5	120							
10	2	142							
11	2.5	156							
12	3	162							
13	3.5	160							
14	4	150							
15	4.5	132							
16	5	106							
17	5.5	72							
18	6	30							
19	6.5	−20							
20	7	−78							
21									
22									

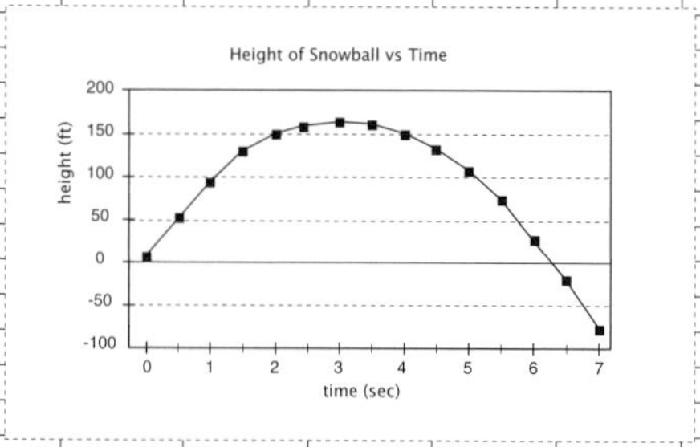

Height of Snowball vs Time

Equation-solving Software Equation-solving software, or *equation solvers,* are a class of software programs that allow you to solve sets of mathematical equations without programming in a conventional computer programming language. An equation solver typically lets you enter equations or formulas "as is," without having to worry about the syntax, order, and grammar of a programming language. The solver then uses a unique method of deriving the solution through the set of equations. This class of software has an assortment of built-in functions, enables you to define your own functions and programs, can display graphical solutions, and provides instantaneous error diagnostics. These tools are also invaluable for documentation of the problem-solving process. Most have built-in word processors with the ability to imbed fully formatted mathematical expressions, Greek symbols, and graphical plots.

Several widely used equation solvers are available to engineering students, including computer-aided design (CAD) programs such as Math-CAD, Matlab, Mathematica, Eureka, Derive, and TKSolver. These packages are similar in their function and use.

Figure 2-8 illustrates an equation solver solution to the snowball problem. This particular solution is generated by the MathCAD program. The equations defining the motion of the snowball are entered as mathematical functions, and the software automatically creates a graph from these functions. A detailed discussion of these tools is beyond the scope of this text.

Figure 2-8
MathCAD solution to the snowball problem

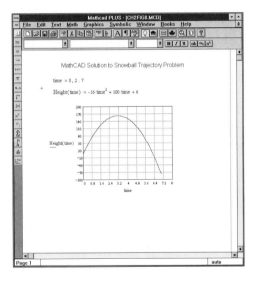

High-Level Programming Languages Spreadsheets, equation solvers, and CAD programs are examples of *application software.* Application software is prewritten software designed to help you use the computer to accomplish a specific task. For example, you can use spreadsheets to perform financial analysis and solve certain types of engineering problems, word processors to create reports and documents, and graphics programs to graphically display information. Choosing the appropriate application software can greatly simplify the fourth step of the problem solving process. Once you've developed a mathematical model defining your particular problem, an appropriate application program will simplify the calculations. For example, a spreadsheet program was used to generate the graph required for a graphical solution to the snowball problem. Although you could also draw the graph by hand, the appropriate software program made the job easier and faster.

Since application programs are designed to solve very specific types of problems, they may not be useful for the particular problem you are trying to solve. To solve some complex problems, you may need to develop an *algorithm* and then translate the algorithm into a computer program. An algorithm is the solution strategy or sequential process required to solve the mathematical model and produce results to your problem. The process of converting the algorithm into instructions the computer can understand is known as *programming*. The result of this activity is called a *computer program*. Since computers are binary information processors, they only "understand" *machine language*, which is a binary string of 0s and 1s. Before the computer can solve a mathematical model for you, the program has to be converted into the binary machine language the computer understands. Programming early computers was a difficult and tedious process, as the mathematical model had to be directly translated into the binary machine language. Fortunately, modern computers use *high-level programming languages* to accomplish this step for you. These languages allow a programmer to communicate with the machine through English-like statements rather than binary strings. Computer languages have a vocabulary and grammar much like human languages. Once a programmer understands the language, it is relatively easy to translate the mathematical model into program statements that instruct the machine how to solve the problem. A *compiler* or *interpreter* translates a high-level language into binary 0s and 1s that the computer can use. Many high-level languages are available today for computer programming. Some languages are designed to handle certain applications better than others. The most popular languages for solving technical problems are FORTRAN, C, BASIC, and Pascal. As an engineer, you will use one or more of these high-level languages to solve technical problems.

Computer-aided Drafting and Design (CADD) Software   You use the previously mentioned computer tools for analysis problems. You use the next tool for design problems. A fundamental component of engineering problem solving and design is the communication of the problem and its solution to others. Until recently, engineers could use two methods to represent their designs and problem solutions: (1) create drawings representing the design and solution on a two-dimensional medium such as paper, and (2) create a three-dimensional physical model or scaled prototype. During the last few years *computer-aided design and drafting (CADD)* software has been replacing both methods as a preferred medium for representing the solution to engineering problems. CADD software creates an electronic representation or model of your problem solution and stores it on a computer system. This software saves you time when you edit and change the drawings.

Computer-aided modeling, as shown in Figure 2-9, implies the creation of a complete representation of the physical object stored in an electronic

Figure 2-9
**Three-dimensional
computer-aided design**

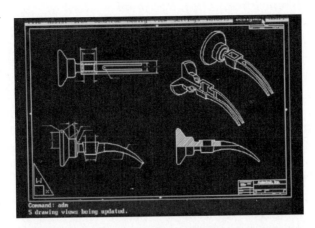

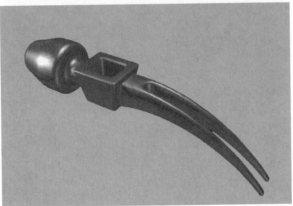

computer database. Because the representation of the object or problem solution is stored electronically in a computer, it is much more convenient to manipulate and evaluate than the actual object. If the computer model accurately represents the physical object and you have the appropriate software available, you can perform any type of analysis on the electronic representation of the actual object. For example, if you design a new type of bicycle frame, you can represent your design electronically with the appropriate three-dimensional computer-aided design software. You can perform an analysis on the computer model to determine the response of your design to various physical conditions. The computer can generate and predict how the bicycle frame will behave in response to physical situations that would damage or even destroy the actual model. You can modify the design electronically until you have a solution that will be strong enough to handle all applied loads and be aesthetically pleasing.

Besides analyzing the solution to an engineering problem, CADD software can document and graphically represent your solution to others. CADD software is an efficient tool to present a graphical solution to engineering problems. The solution to many engineering problems is an *iterative process*, where the original solution is repeatedly modified and refined. Before CADD, such an iterative process required the engineer to continually modify and redraw the problem solution. With an electronic model of your problem solution, you can quickly perform modifications using the appropriate CADD software tools.

Electronic Information Retrieval   A critical step in the problem-solving process is step 2, collecting information. Much of an engineer's time is spent searching for information or transferring information to someone else. Although engineers are primarily problem solvers, they seldom have all the information they need to completely solve a problem. During the last few years, we've witnessed an explosion of information. Engineers have traditionally found and collected information by searching through printed resources. Your library will have many different printed resources that contain information pertinent to engineering problems.

Until the late 1980s the conventional medium for information storage, retrieval, and transfer was printed material. As the communication infrastructure developed in the United States, information flowed faster and in greater quantities than ever before. No longer were printed books, journals, and indexes sufficient to handle the explosion of information. Electronic bulletin boards, electronic mail, and computer databases began to proliferate as computer networks of all sorts stretched across the country. While electronic information storage and retrieval has not yet completely replaced printed sources, it is a vital source of the most current information regarding technical problems.

The most widely used and fastest growing resource for electronic information transfer is the *Internet*. The Internet is a global network connecting tens of thousands of separate computer networks. The Internet is also a community of people who use and develop those networks and a collection of resources that can be reached from those networks. By the mid-1990s the Internet connected 30,000 smaller networks spanning 45 different countries. Over 1 million people use the Internet every day.

The Internet provides you with instant access to 30,000 computer networks and 2.5 million users. With an Internet connection you have several valuable resources and information sources for engineering design and problem solving:

- *Electronic mail (E-mail).* Anyone who can access the Internet either through a university or home computer has an electronic mailbox. With this tool you can send and receive electronic mail, or *E-mail,* almost instantly with over 20 million other mailboxes around the world. E-mail can be forwarded, saved, or printed. You can also set up mailing lists, allowing the same message to be sent simultaneously to multiple addresses. Data files can be attached to your mail message, thereby allowing the exchange of electronic information.
- *Remote log-on.* With access to the Internet, you have the capability of logging onto any computer on the network. This means that you can work at home, or even overseas, and access another computer as if you were on site.
- *File transfer.* In 1994 millions of files were on-line, representing hundreds of gigabytes of data. These files are available to anyone on the network. These files might be computer programs, graphics files, or text files, to name a few.
- *Electronic journals.* Thousands of electronic journals available for subscription cover a multitude of special topics, touching nearly every engineering discipline. They are usually available at no charge by

subscribing electronically. You can even get a current weather report on any location in the world!

- *Electronic bulletin boards.* Available on the Internet, electronic bulletin boards cover various special topics related to engineering. For example, you can log onto a bulletin board specifically covering C++ programming. Anyone on the network (millions of other people) can contact this bulletin board to post questions to others or submit an answer to a question. Lively, fast-moving discussions can arise on the various special interest bulletin boards. If you need information about your specific engineering problem, there is a good chance you can get the answer on the appropriate bulletin board.
- *University library catalogues.* Most major universities now have their catalogues available on the Internet. You can do a remote search for a specific title, author, or subject in any one of these libraries. Generally, the books or journals are available through interlibrary loan to your university library. Many libraries even contain electronic copies of technical journal articles, which can be faxed or transmitted electronically right to your home.

This list is just a glimpse of what is available on the Internet. Electronic networks are becoming a key part of science and engineering. Engineers can now use computers to share ideas and information with other scientists and engineers, to collaborate with colleagues, and exchange engineering designs and data over the network instantaneously.

Try It

This chapter began with a description of the complex technical problems engineers had to solve before they could design and build an internal combustion engine. One of these problems was to determine the pressure inside the piston-cylinder assembly during the operation of the engine and the corresponding force exerted by the piston. Pressure is defined as the force per unit area exerted by a fluid on some surface. This quantity is important because the high pressure generated inside the engine cylinder produces the force on the piston surface that moves it down the cylinder. The motion of the crankshaft and the work done by the engine are therefore directly related to the pressure inside the cylinder. Figure 2-10 shows a cutaway view of an automobile engine illustrating the workings of the piston-cylinder arrangement.

The internal combustion engine shown in Figure 2-10 has a minimum volume at the top of the cylinder equal to 1.0 in^3. The diameter of the piston is 3.0 in. The conditions inside the engine have been measured immediately *before* combustion occurs, and the pressure was found to be $P_1 = 100$ pounds per in^2 (psi). Immediately after combustion, the pressure inside the engine is measured to be $P_z = 600$ psi. For these given conditions, determine the maximum force exerted by the piston after combustion.

To solve this problem, you need to know the relationship between several variables in this system:

- The relationship between force and pressure
- The relationship between surface area and geometry in the engine.

Figure 2-10
Cutaway view of a typical automobile engine

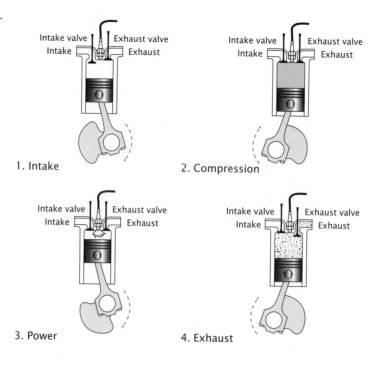

1. Intake

2. Compression

3. Power

4. Exhaust

SUMMARY

This chapter introduced you to a general five-step problem-solving process you can use to solve analysis problems. You can apply this process to technical problems that engineers solve. The five steps are

1. Define the problem clearly.
2. Collect all the information pertinent to the problem.
3. Generate a solution to the problem.
4. Refine and implement the solution.
5. Verify and test the solution.

This chapter also defined the distinction between analysis problems and design problems. Modern tools that engineers use for problem solving were also presented in this chapter, including a brief introduction to computer tools.

Key Words

algorithm
analysis
application software
approximation
assumption
cadence
cell
compiler
computer-aided design
  and drafting (CADD)
computer program

conservation of energy
derived unit
dimensional consistency
electronic spreadsheet
E-mail
engineering paper
equation solver
estimation
fundamental unit
graphical solution
high-level programming language

Internet

interpreter

iterative process

kinetic energy

machine language

potential energy

pressure

programming

template

unit

Exercises

1. The bending moment of shaft supported by bearings is

$$M = \frac{wL^2}{12}$$

where M is the bending moment in inch-lbf, L is the span between bearings in inches, and w is the weight of a unit length of shaft in lbf/inch. For a shaft with a weight $w = 20$ lbf/inch, determine the bending moment for a 6-inch shaft, 12-inch shaft, and 24-inch shaft. Create a plot of M versus L. Verify that your solution is reasonable and realistic.

2. The following temperature data was measured by a weather balloon as it flew from an elevation of 0 meters to 5000 meters above sea level. If available, use a spreadsheet to plot the temperature as a function of elevation and determine what the temperature is at an elevation of 3200 meters above sea level.

Altitude (meters)	Temperature (°C)
0	20
1000	12
2000	7
3000	0
4000	-5
5000	-10

3. An electric utility company charges its customers 8 cents per kilowatt hour (kwh) of electricity usage up to 1000 kwh. After 1000 kwh the rate goes up to 12 cents per kwh. A family averages 2 kw of usage for the first 21 days of the month. During the last 10 days of the month, the family is on vacation and their average rate of electric usage drops to 1.0 kW. What is their utility bill for the month? Estimate the answer by hand, and then use a spreadsheet or equation solver to find the solution.

4. An electric power plant burns coal at the rate of 1500 kg/hr. The coal has a sulfur content of 2 percent. Before putting an emission control system on the exhaust of the plant, all of this sulfur ended up being emitted into the environment. (a) How many kg of sulfur were emitted each day by the plant? (b) After modifying the plant and installing a clean-up system, 80 percent of the sulfur is captured before leaving

the exhaust stacks of the plant. How many kg of sulfur are emitted each day after the modification? Verify that your solution is reasonable and realistic.

5. The work output of an automobile engine is measured at different speeds. The following experimental data is obtained for work (horsepower) as a function of engine speed (RPM). If available, use a spreadsheet to create a plot of the data, and determine the work output of the engine at speeds of 1500 rpm and 3200 rpm.

Engine Speed (rpm)	Engine Work Output (hp)
0	0
1000	65
2000	120
3000	140
4000	105
5000	60

6. An automobile has tires with a diameter, $d = 25$ inches. The tires on this car turn at a rate of 50 radians/minute. Estimate the speed of the car in miles per hour. Then calculate the results using a spreadsheet.

7. The automobile in problem 6 maintains this constant speed for a period of 10 minutes. How many miles has the car traveled? Is the answer reasonable and realistic?

8. The automobile in problem 6 accelerates at a constant rate so the tires turn from 0 to 50 radians/minute over the time interval of 3 minutes. How far did the car travel during this 3 minute time interval? (Hint: Plot a graph of speed as a function of time from 0 to 3 minutes. The distance is the area under the curve). First estimate the answer, and then calculate the result using a spreadsheet or equation solver.

9. A rectangular room 19.2 feet by 29.6 feet has an 11-foot by 20-foot rung placed symmetrically on the floor. How many square yards of floor will be exposed? Is the answer reasonable and realistic?

10. The speed of a ship traveling at the rate of 10 km/hr uniformly slows down to 5 km/hr in a distance of 0.5 km. Assume that the rate of deceleration remains constant. (a) Plot the ship's speed as a function of time. (b) How long in minutes will it take the ship to come to a rest? (c) How many km will it have traveled from the point where its speed was 10 km/hr? If available, use a spreadsheet to calculate the results.

11. An automobile climbs the ramp of a parking garage that has a slope of 1 foot vertical for each 4 feet of horizontal distance. The second level of the garage is 12 feet above the ground level. The car has tires that are 30 inches in diameter. How many revolutions do the wheels make in climbing the ramp? Is the answer reasonable and realistic?

12. A helical spring is 10 cm long. Its mean diameter is 5 cm and the pitch is 1 cm (pitch is the length of one coil of the spring). Estimate the length of wire in the spring. Then calculate the answer by hand or using a spreadsheet.

13. The density of steel is 490 lb per ft^3. What is the weight of a piece of steel having the shape of an equilateral triangle with sides 3 feet long? The plate is 5/8-inch thick and has two holes with a 3-inch diameter drilled through it. Use a spreadsheet or equation solving software to solve the problem.

14. How many square meters of roofing will be required to roof a house that is 10 meters wide by 15 meters long. The roof is 2.5 meters higher at the ridge than at the eaves. Is the answer reasonable and realistic?

15. Two pulleys 20 inches and 30 inches in diameter respectively are connected by a belt whose speed is 2500 feet per minute. What is the rpm of each pulley? Is the answer reasonable and realistic?

3

Engineering Design: A Creative Process

Jet Air Travel — Today you can travel around the world in a matter of hours. Yet only a hundred years ago, transcontinental travel was long and arduous, and it took weeks or months to cross the Atlantic Ocean. This dramatic change didn't occur over night. The modern jet was developed by numerous engineers over the last 75 years. First, the Wright brothers applied their knowledge of physics and mechanics to create a machine that could fly under its own power. The succeeding generations of engineers continuously modified and perfected early aircraft to create the machines we now take for granted. The process of applying technical knowledge to create new machines or processes or to modify existing ones to meet new needs is called Design.

INTRODUCTION

If you take a moment to observe your surroundings, you will see examples of technological creativity. The physical objects you see, whether they be telephones, automobiles, bicycles, or electric appliances, all came into being through the creative application of technology. These everyday inventions did not miraculously appear but originated in the minds of human beings and took time to develop. Engineering is the creative process of turning abstract ideas into physical representations (products or systems). What distinguishes engineers from painters, poets, or sculptors is that engineers apply their creative energies to producing technical devices or systems that meet human needs. This creative act is called design.

3-1 WHAT ARE ENGINEERING DESIGNS?

Engineering designs can be classified as inventions—devices or systems that are created by human effort and did not exist before or are improvements over existing devices or systems. Inventions, or designs, do not suddenly appear from nowhere. They are the result of bringing together technologies to meet human needs or to solve problems. Sometimes a design is the result of someone trying to do a task more quickly or efficiently. Design activity occurs over a period of time and requires a step by step methodology.

Chapter 1 described engineers primarily as problem solvers. What distinguishes design from other types of problem solving is the nature of both the problem and the solution. Design problems are open ended in nature, which means they have more than one correct solution. The result or solution to a design problem is a system that possesses specified properties.

Design problems are usually more vaguely defined than analysis problems. Recall the snowball problem from Chapter 2. The original problem statement asked you to determine the maximum height of the snowball given an initial velocity and release height. If you change the problem statement to read, "Design a device to launch a 1-pound snowball to a height of at least 160 feet," this analysis problem becomes a design problem. The solution to the design problem is a system having specified properties (able to launch a snowball 160 feet), whereas the solution to the analysis problem consisted of the properties of a given system (the height of the snowball). The solution to a design problem is therefore open ended, since there are many possible devices that can launch a snowball to a given height. The original problem had a single solution: the maximum height of the snowball, determined from the specified initial conditions.

Solving design problems is often an iterative process: As the solution to a design problem evolves, you find yourself continually refining the design. While implementing the solution to a design problem, you may discover that the solution you've developed is unsafe, too expensive, or will not work. You then "go back to the drawing board" and modify the solution until it meets your requirements. For example, the Wright brothers' airplane did not fly perfectly the first time. They began a program for building an airplane by first conducting tests with kites and then gliders. Before attempting powered flight, they solved the essential problems of controlling a plane's motion in rising, descending, and turning. They didn't construct a powered plane until after making more than 700 successful glider flights. Design activity is therefore cyclic or iterative in nature, whereas analysis problem solving is primarily sequential.

The solution to a design problem does not suddenly appear in a vacuum. A good solution requires a methodology or process. There are probably as many processes of design as there are engineers. Therefore, this book does not present a rigid "cookbook" approach to design but presents a general application of the five-step problem-solving methodology introduced in Chapter 2 for design problems. Not every design problem will fit this approach exactly. The process described here is general, and you can adapt it to the particular problem you are trying to solve.

3-2 THE DESIGN PROCESS

The basic five-step problem-solving process presented in Chapter 2 also works for design problems. While the terminology and methodology are the same for both types of problem-solving activities, the actual application of each step differs. Performing each step sequentially yields the solution to a closed-ended analysis problem. Since design problems are usually defined more vaguely and have a multitude of correct answers, the process may require backtracking and iteration. Solving a design problem is a *contingent process*—the solution is subject to unforeseen complications and changes as it develops. Until the Wright brothers actually built and tested their early gliders, they did not know the problems and difficulties they would face controlling a powered plane.

The five steps used for solving design problems are

1. Define the problem.
2. Gather pertinent information.
3. Generate multiple solutions.
4. Analyze and select a solution.
5. Test and implement the solution.

Notice that steps 3, 4, and 5 are slightly modified from the corresponding steps in solving analysis problems. These changes are necessary because of the open-ended nature of design problems. Since the design process is producing multiple answers to the problem definition, steps 3 and 4 have been modified to generate, analyze, and select multiple solutions. We discuss these steps later in this chapter.

When solving a design problem, you may find at any point in the process that you need to go back to a previous step. The solution you chose may prove unworkable for any number of reasons and may require redefining the problem, collecting more information, or generating different solutions. This continuous iterative process is represented in Figure 3-1.

Step 1: Define the Problem

You need to begin the solution to a design problem with a clear, unambiguous definition of the problem. Unlike an analysis problem, a design problem often begins as a vague, abstract idea in the mind of the designer.

Figure 3-1
The design process

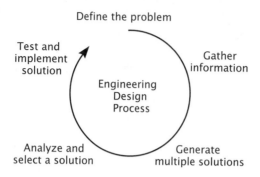

Creating a clear definition of a design problem is more difficult than defining an analysis problem. The definition of a design problem may evolve through a series of steps or processes as you develop a more complete understanding of the problem.

Identify and Establish the Need   Engineering design activity always occurs in response to a human need. Before you can develop a problem definition statement for a design problem, you need to recognize the need for a new product, system, or machine. Thomas Newcomen saw the need for a machine to pump the water from the bottom of coal mines in England. Recognizing this human need provided him the stimulus for designing the first steam engine in 1712. Before engineers can clearly define a design problem, they must see and understand this need.

Although engineers are generally involved in defining the problem, they may not be the ones who initially recognize the need. In private industry, market forces generally establish the need for a new design. A company's survival depends on producing a product that people will buy and can be manufactured and sold at a profit. Ultimately, consumers establish a need, because they will purchase and use a product that they perceive as meeting a need for comfort, health, recreation, transportation, shelter, and so on. Likewise, the citizens of a government decide whether they need safe drinking water, roads and highways, libraries, schools, fire protection, and so on.

The *perceived* need, however, may not be the *real* need. Before you delve into the details of producing a solution, you need to make sure you have enough information to generate a clear, unambiguous problem definition that addresses the real need. The following example illustrates the importance of understanding the need before attempting a solution.

EXAMPLE 3-1

Automobile Airbag Inflation: How Not to Solve a Problem

A company that manufactures automobile airbags has a problem with an unacceptably high rate of failure in the inflation of the bag. During testing, 10 percent of the bags do not fully inflate. An engineer is assigned the job of solving the problem.

At first the engineer defines the problem as a failure in the materials and construction of the inflation device. The engineer begins to solve this problem by producing a more robust inflation device. After considerable

Figure 3-2
Newcomen steam engine and water pump, 1712

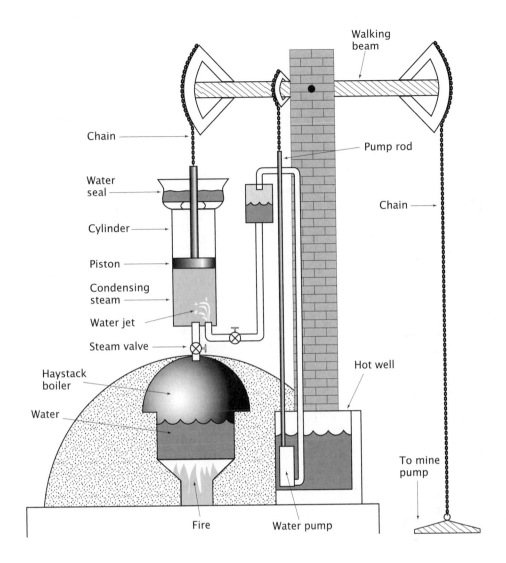

effort, the engineer discovers that improving the inflation device does not change the failure rate in the bags. Eventually, this engineer re-examines the initial definition of the problem. The company investigates the airbag inflation problem further and discovers that a high degree of variability in the tightness of folds is responsible for the failure of some bags to inflate. At present the bags are folded and packed by people on an assembly line.

With a more complete understanding of the need, the engineer redefines the problem as one of increasing the consistency in tightness of the folds in the bags. The final solution to this problem is a machine that automatically folds the bags.

· ·

Often the apparent need is not the real need. A common tendency is to begin generating a solution to an apparent problem without understanding the problem. This approach is exactly the wrong way to begin solving a problem such as this. You would be generating solutions to a problem that has never been defined.

People have a natural tendency to attack the current solution to a problem rather than the problem itself. Attacking a current solution may eliminate inadequacies but will not produce a creative and innovative solution. For example, the engineer at the airbag company could have only looked at the current method for folding airbags—using humans on an assembly line. The engineer might have solved the problem with inconsistent tightness by modifying the assembly line procedure. However, the final solution to the problem proved to be more cost effective and reliable, in addition to producing a superior consistency in the tightness of the folds.

Develop a Problem Statement   The first step in the problem-solving process, therefore, is to formulate the problem in clear and unambiguous terms. Defining the problem is not the same as recognizing a need. The problem definition statement results from first identifying a need. The engineer at the airbag company responded to a need to reduce the number of airbag inflation failures. He made a mistake, however, in not formulating a clear definition of the problem before generating a solution. Once a need has been established, engineers define that need in terms of an engineering design problem statement. To reach a clear definition, they collect data, run experiments, and perform computations that allow that need to be expressed as part of an engineering problem-solving process.

Consider for example the statement "Design a better mousetrap." This statement is not an adequate problem definition for an engineering design problem. It expresses a vague dissatisfaction with existing mousetraps and therefore establishes a need. An engineer would take this statement of need and conduct further research to identify what was lacking in existing mousetrap designs. After further investigation the engineer may discover that existing mousetraps are inadequate because they don't provide protection from the deadly hantavirus carried by mice. Therefore, a better mousetrap may be one that is sanitary and does not expose human beings to the hantavirus. From this need, the problem definition is modified to read, "Design a mousetrap that allows for the sanitary disposal of the trapped mouse, minimizing human exposure to the hantavirus."

The problem statement should specifically address the real need yet be broad enough not to preclude certain solutions. A broad definition of the problem allows you to look at a wide range of alternative solutions before you focus on a specific solution. The temptation at this point in the design process is to develop a preconceived mental "picture" of the problem solution. For example, you could define the better mousetrap problem as "Design a mousetrap that sprays the trapped mouse with disinfectant." This statement is clear and specific, but it is also too narrow. It excludes many potentially innovative solutions. If you focus on a specific picture or idea for solving the problem at this stage of the design process, you may never discover the truly innovative solutions to the problem. A problem statement should be concise and flexible enough to allow for creative solutions.

Here is one possible problem definition statement for our better mousetrap problem:

> **Problem Definition Statement: A Better Mousetrap**
> Certain rodents such as the common mouse are carriers and transmitters of an often fatal virus, the hantavirus. Conventional mousetraps expose people to this virus as they handle the trap and dispose of the mouse. Design a mousetrap that allows a person to trap and dispose of a mouse without being exposed to any bacterial or viral agents being carried on the mouse.

Establish Criteria for Success   *Criteria for success* are the specifications a design solution must meet or the attributes it must possess to be considered successful. You should include criteria in the problem statement to provide direction toward the solution. At this point in the design process, the criteria are preliminary. As the design solution develops, you will most likely find that the initial criteria need to be redefined or modified. Preliminary criteria must not be too specific so they allow flexibility through the design process.

The criteria that apply to a particular design problem are based on your background knowledge and the research that you've conducted. Since each problem or project is unique, the desirable attributes, or criteria, of the solution are also unique. Some criteria are unimportant to the success of the design. The list of criteria is developed by the *design team*. The design team is made up of people from various engineering backgrounds who have expertise pertinent to the problem. This team may also include people from backgrounds other than engineering, such as managers, scientists, and technicians. The design team must evaluate each criterion and decide if it is applicable to the design effort. Later in the design process, value judgments must be applied to the list of criteria. Therefore, it makes little sense to include those criteria that will be of relatively low priority in the evaluation of design solutions. For example, if you are designing a critical life support system, you would not include the criterion of "must be minimum cost," because cost is not an important factor in evaluating this design.

The following is a list of preliminary criteria for a better mousetrap design. This list would be included in the problem definition statement.

> **Criteria for Success of a Better Mousetrap**
> - The design must be low cost.
> - The design should be safe, particularly with small children.
> - The design should not be detrimental to the environment.
> - The design should be aesthetically pleasing.
> - The design should be simple to operate, with minimum human effort.
> - The design must be disposable (you don't reuse the trap).
> - The design should not cause undue pain and suffering for the mouse.

Try It   Develop a problem definition statement and a list of criteria for success for the design of an entrance/exit door in a residential home for a household pet.

What If

The user of the better mousetrap would like an alarm that signals when a mouse is trapped. Redo the problem definition statement and criteria for success list for the better mousetrap to reflect this modification.

Step 2: Gather Pertinent Information

Before you can go further in the design process, you need to collect all the information available that relates to the problem. Novice designers will quickly skip over this step and proceed to the generation of alternative solutions. You will find, however, that effort spent searching for information about your problem will pay big dividends later in the design process. Gathering pertinent information can reveal facts about the problem that result in a redefinition of the problem. You may discover mistakes and false starts made by other designers.

Information gathering for most design problems begins with asking the following questions. If the problem addresses a need that is new, then there are no existing solutions to the problems, so obviously some of the questions would not be asked.

- What are the existing solutions to the problem?
- What is wrong with the way the problem is currently being solved?
- What is right about the way the problem is currently being solved?
- What companies manufacture the existing solution to the problem?
- What are the economic factors governing the solution?
- How much will people pay for a solution to the problem?
- What other factors are important to the problem solution (such as safety, aesthetics, environmental issues, and color)?

Search for Information Resources   As an engineering student in the 1990s you have many more sources of information available to you than engineers did only 20 years ago. This section discusses some of the most current resources available, but because our world is witnessing an information explosion, by the time you read this many more resources will be available that are not mentioned here.

In 1989 researchers at the University of Utah shook the scientific community with their discovery of cold fusion. They claimed to have discovered a way to release the power of the atom in a test tube, with no radioactive waste, no danger, and almost no cost. The popular press and the scientific community were asking the question, Could it be true? Scientists and engineers around the world wanted to know the details and wanted to know what other scientists had discovered when they tried to re-create the experiment. These researchers needed the information instantly, and traditional publications had a delay time of months or even years. Many of these researchers turned to the latest methods of information transfer—E-mail and the fax machine. Experimental results, arguments, and controversy spread around the world at the speed of light. Electronic information transfer can reach a much larger audience significantly quicker than traditional publications of scientific theory and results. Due to the rapid transfer of information, cold fusion was quickly debunked.

Traditional publications are still an essential source of information to engineers and scientists. However, electronic information transfer and retrieval are quickly becoming a standard source for engineers and scientists. When you begin a search for information relating to a design problem, you must be prepared to go to many different sources.

The library is still the primary source of information for an engineering student. Your success as an engineer and student will be enhanced if you are able to use the library effectively. For specific help on using your library, you should consult the library staff at your university; they probably offer courses or seminars on library usage. Some of the common resources available at a university library are discussed below.

- **Scientific encyclopedias and technical handbooks.** These sources are a good place to start when you are investigating an area or problem that is new to you. An encyclopedia or handbook provides a brief general overview by an authority in a particular field and includes references for more detailed information. The McGraw-Hill *Encyclopedia of Science and Technology* covers all scientific fields. Technical handbooks, such as the *Electrical Engineers Handbook* or *Mark's Handbook of Mechanical Engineering*, cover various fields such as chemical, civil, electrical, or mechanical engineering. The information contained in these handbooks is presented in a very concise form and can be good starting points for an in depth search.
- **Card catalog.** Card catalogs give a listing of all the sources available at your library. They are categorized by subject matter, author, or title. The subject card gives a brief summary of the book's content, including the title, author, publisher, copyright date, and total number of pages. The call number tells you where to locate the book in your library. Recently, many libraries have begun to file their card catalogs electronically. Instead of going to a physical stack of cards, you use a computer terminal to search for a book by subject, author, or title.
- **Indexes.** *Indexes* categorize current works in various disciplines. They list the subject, title, and author of recent articles in technical and trade journals under various subject headings. Some indexes include brief abstracts of articles. Most indexes are updated monthly, so a complete search through an index may be tedious. A familiar index to scientists and engineers is the *Index of Applied Science and Technology*. It lists articles from 335 journals and is updated monthly. The *Engineering Index* is another popular index for engineers. It selects articles from approximately 2700 journals and periodic publications and includes an abstract of each article.
- **Electronic searches.** Computer networks and electronic databases are quickly replacing printed sources of information. Most university libraries and many public libraries have now catalogued their holdings in electronic databases. These are electronic versions of card catalogs. To search for a subject, title, or author, you use a terminal located in the library, or you can connect to the database through a computer network. Your library staff can give you instructions on how to use the system at your university. The Internet, which was discussed in Chapter 2, now links most major university library catalogs. With an Internet con-

nection you have access to hundreds of libraries around the world. You can search for a book or journal by subject, title, author, or key words. If your local library does not have the material you need, you can usually order it through interlibrary loan. For a current list of the many indexes available on the Internet, you need to see your librarian. The list changes almost daily, so any list printed in this module would be out of date by the time you read it.

 Try It   Choose an automobile accessory and list five potential sources of information that could support its design.

Record the Results   You should record a bibliography as you search and review sources of information. You will prepare a written report later in the design process, and the information you collect will be cited in this report. You may be frustrated at later stages in the design process, when you are trying to locate information that you've already seen but can't find. Record the information in a form that makes it easy to cite as a reference in your written reports. Be sure to include the title, author, date, journal name and number, call number if it's a book, and a brief synopsis of the article. Some people record their bibliography on index cards, others find it useful to keep records in a bound notebook. Record any conversations with vendors or other engineers in your notebook, including each person's name and the date and pertinent details of the conversation.

 What If   What additional sources of information might you need to add an alarm system to the design of a better mousetrap?

Step 3: Generate Multiple Solutions

The next step in the design process begins with *creativity*—generating new ideas that may solve the problem. Creativity is much more than just a systematic application of rules and theory to solve a technical problem. Before you examine creative problem solving more closely, take a look at the definition of creativity that Edward and Monika Lumsdaine give in their book, Creative Problem Solving[1]:

> Creativity is playing with imagination and possibilities,
> leading to new and meaningful connections and outcomes
> while interacting with ideas, people and the environment.

---

1. Edward and Monica Lumsdaine, *Creative Problem Solving*, 2d Ed., New York: McGraw-Hill, 1995.

You start with existing solutions to the problem and then tear them apart—find out what's wrong with those solutions and focus on how to improve their weaknesses. Consciously combine new ideas, tools, and methods to produce a totally unique solution to the problem. This process is called *synthesis*. Casey Golden, age 13, did this when he invented the BIOtee[2]. Casey noticed that discarded and broken wooden golf tees littered golf courses, damaging the blades and tires of lawn mowers. He decided to design a new biodegradable tee. After experimenting with different mixtures, he devised a recipe made of recycled paper fiber and food byproducts coated with a water-soluble film. When the film is broken, moisture in the ground breaks down the tee within 24 hours. As a result of his creative efforts, Casey's family started a company to manufacture BIOtees and now has orders for six million per year.

Psychological research has found no correlation between intelligence and creativity. People are creative because they make a conscious effort to think and act creatively. Everybody has the potential to be creative. Creativity begins with a decision to take risks. Listed below are a few characteristics of creative people. These are not rigid rules to be followed to experience creativity. You can improve your creative ability by choosing to develop these characteristics in yourself.

- **Curiosity and tolerance of the unknown.** Creative people have a positive curiosity of the unknown. They are not afraid of what they don't understand.
- **Openness to new experiences.** Creative people have a healthy and positive attitude toward new experiences.
- **Willingness to take risks.** Creative people are not afraid to take risks and try new experiences or ideas, knowing that they may be misunderstood and criticized by others. They are self-confident and not afraid to fail.
- **Ability to observe details and see the "whole picture."** Creative people notice and observe details relating to the problem, but they also can step back and see the bigger picture.
- **No fear of problems.** Creative people are not afraid to tackle complex problems, and they even search for problems to solve. They seek solutions to problems with their own abilities and experience if possible. They have the attitude of "if you want something done, you'd better do it yourself."
- **Ability to concentrate and focus on the problem until it's solved.** Creative people can set goals and stick to them until they're reached. They focus on a problem and do not give up until the problem is solved. They have persistence and tenacity.

Strategies for Generating Creative Solutions Creative solutions to engineering design problems do not magically appear. Ideas are generated when people are free to take risks and make mistakes. When Thomas Edison was searching for the right material to make an electric light filament,

---

2. "Kids in Business," *National Geographic World*, January 1993, p. 23.

he tried 2000 different materials before he found one that worked. We tend to forget the 1999 failures when thinking about the success of his invention. Likewise, the members of a design team have to function in a nonthreatening environment, where risk-taking is encouraged. Brainstorming and sketchstorming are proven techniques that enhance the generation of creative ideas.

Brainstorming is a technique of generating many ideas with the hope that a few good ideas will develop into something workable. Brainstorming is a group activity; it only works when the members of the group develop *synergism*—the recognition that the product of the whole group is greater than the product of the individual members. This works because the ideas of one member of the group stimulate creativity in another member. Usually brainstorming occurs over a short period, 30 minutes to an hour. Successful sessions generally involve at least three people, no more than fifteen. Six to eight people in a group is an optimum size for a successful session. Here are a few guidelines for a brainstorming session:

- **Accept everything.** Withhold criticism or evaluation of ideas. Everybody in the group is treated as an equal. No one considers one person's ideas better than someone else's. All members of the group withhold judgment until a later stage.
- **Welcome the outlandish.** New ideas are born only when freedom to hatch them exists. There are no "dumb" ideas. The unworkable solutions will be evaluated and thrown out later.
- **Stress the importance of quantity.** The group is encouraged to generate as many ideas as possible. As the number of ideas increases, so does the probability that really good ideas will surface.
- **Build on and combine old ideas.** Brainstorm ways that existing solutions can be modified or combined into a new solution.
- **Record everything said for later evaluation.** One person in the group is the designated record keeper. That one person writes down quick sketches of all ideas. These sketches will be reviewed and refined at a later stage in the design process. This is illustrated in the example at the end of the chapter.

Sketchstorming is the engineer's response to brainstorming. Sketchstorming is the visual creation and recording of ideas. Since solutions to engineering problems typically come in visual rather than verbal images, it is important to record these ideas onto paper in sketch form. Sketching enhances the refinement of these ideas at a later stage. Sketches are not detailed drawings of your ideas. They are quick, two-dimensional representations of what your mind is seeing. Sketching does not require artistic talent. Many great inventors and scientists such as Leonardo da Vinci and Thomas Edison kept a visual record of their ideas. Sketching ideas quickly on paper allows you to store the visual image, modify the idea, and add details to the design later. Figure 3-3 shows some of Thomas Edison's sketches of a mechanism to advance paper tape in a printing head.

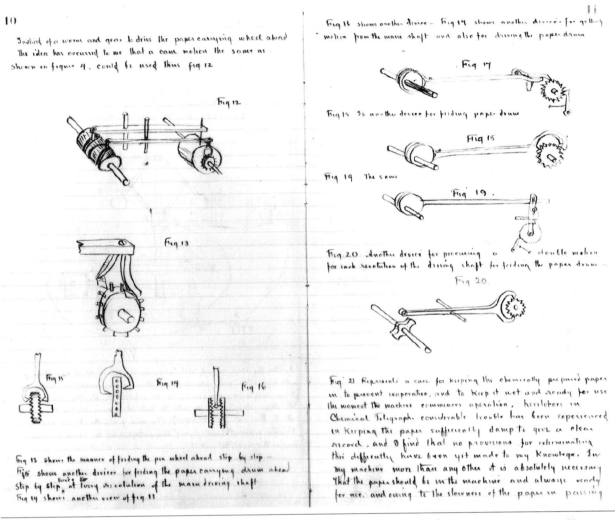

Figure 3-3   Thomas Edison's sketches of a mechanism used in his inventions

Try It

Form a design team and practice a brainstorming exercise to generate alternative-solution ideas for the design of a residential entrance/exit door for a household pet. Use sketchstorming to record these ideas.

Step 4: Analyze and Select a Solution

Once you've conceived alternative solutions to your design problem, you need to analyze those solutions and then decide which solution is best suited for implementation. Analysis is the evaluation of the proposed designs. You apply your technical knowledge to the proposed solutions and use the results to decide which solution to carry out. You will cover design analysis in more depth when you get into upper-level engineering courses.

At this step in the design process, you must consider the results of your design analysis. This is a highly subjective step and should be made by a group of experienced people. This section introduces a systematic methodology you can use to evaluate alternative designs and assist in making a decision.

Analysis of Design Solutions   Before deciding which design solution to implement, you need to analyze each alternative solution against the selection criteria defined in step 1. You should perform several types of analysis on each design. Every design problem is unique and requires different types of analysis, but the following types of analysis apply to most problems:

- Functional analysis
- Ergonomics
- Product safety and liability
- Economic and market analysis
- Strength and mechanical analysis

The following paragraphs provide details of each type of analysis

Functional analysis   determines whether the given design solution will function the way it should. Functional analysis is fundamental to the evaluation and success of all designs. A design solution that does not function properly is a failure even if it meets all other criteria. Consider for example the invention of the ballpoint pen. This common instrument was first invented and manufactured during World War II. The ballpoint pen was supposed to solve the problems of refilling and messiness inherent to the fountain pen. Unfortunately, this new design had never been evaluated for functionality. The early pens depended on gravity for the ink to flow to the roller ball. This meant that the pens only worked in a vertical upright position, and the ink flow was inconsistent: Sometimes it flowed too heavily, leaving smudgy blotches on the paper; other times the flow was too light and the markings were unreadable. The first ballpoint pens tended to leak around the ball, ruining people's clothes. An elastic ink, developed in 1949, allowed the ink to flow over the ball through smooth capillary action. Not until the 1950s did the ballpoint pen finally become a practical writing instrument, thanks to proper ink and engineering. Economy, appearance, durability, and marketability of a design are unimportant if the product does not function properly.

Ergonomics   is the human factor in engineering. It is the study of how people interact with machines. Most products have to work with people in some manner. People occupy a space in or around the design, and they may provide a source of power or control or act as a sensor for the design. For example, people sense if an automobile air-conditioning system is maintaining a comfortable temperature inside the car. These factors form the basis for human factors, or ergonomics, of a design.

A design solution can be considered successful if the design fits the people using it. The handle of a power tool must fit the hand of everybody using it. The tool must not be too heavy or cumbersome to be manipulated by all sizes of people using the tool. The geometric properties of people—their weight, height, reach, circumference, and so on—are called *anthropometric data.* The difficulty in designing for ergonomics is the abundance of anthropometric data. The military has collected and evaluated the distribution of human beings and published this information in military standard tables. A successful design needs to be evaluated and analyzed against the distribution of geometry of the people using it. Figure 3-4 shows the geometry of typical adult males and females for the general population. Since people come in different sizes and shapes, these data are used by design engineers to assure that their design fits the user. A good design will be adjustable enough to fit 95 percent of the people who will use it.

Product Safety and Liability   Because litigation has become common in today's environment, design engineers must be familiar with the issues of safety and liability. *Liability* refers to the manufacturer of a machine or product being liable, or financially responsible, for any injury or damage resulting from the use of an unsafe product. The primary consideration for *safety* in product design is to assure that the use of the design does not cause injury to humans. Safety and product liability issues, however, can also extend beyond human injury to include property damage and environmental damage from the use of your design. Neglect of any of these factors may result in a dangerous situation.

The only way to assure that your design will not cause injury or loss is to design safety into the product. You can design a safe product in three ways. The first method is to design safety directly into the product. Ask yourself, "Is there any probability of injury during the normal use and during failure of your design?" For example, modern downhill ski bindings use a spring-loaded brake that brakes the ski automatically when the ski disengages from the skier's boot. Older ski bindings used an elastic cable attached to the skier's ankle, but this had a tendency to disconnect during a severe fall.

Inherent safety is impossible to design into some products, such as rotating machinery and vehicles. In such cases you use the second method of designing for safety: You include adequate protection for users of the product. Protection devices include safety shields placed around moving and rotating parts, crash protective structures used in vehicles, and "kill" switches that automatically turn a machine off (or on) if there is potential for human injury. For example, new lawnmowers generally include a protective shield covering the grass outlet and include a kill switch that turns the motor off when the operator releases the handle.

The third method to design for safety is the use of warning labels describing inherent dangers in the product. Warning labels are the weakest way to implement safety in a design. In most cases, however, a warning label will not protect you from liability. Protective shields or other devices must be included in the design.

A product liability suit may be the result of a personal injury due to the operation of a particular product. The manufacturer and designer of a device can be found liable to compensate a worker for losses incurred during the operation or use of their product. During a product liability trial, the plaintiff attempts to show that the designer and manufacturer of a product are negligent in allowing the product to be put on the market.

Figure 3-4
Anthropometric data

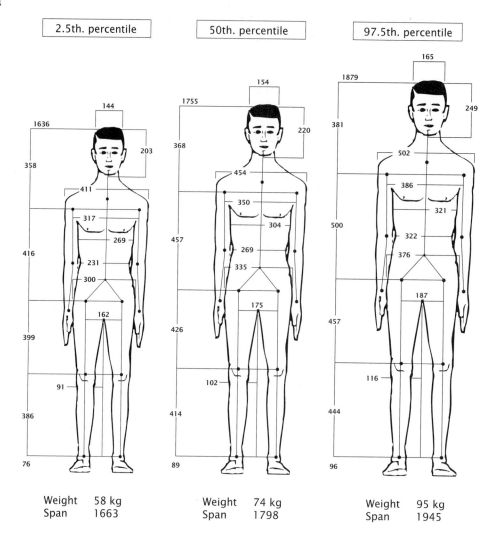

Standing adult male - including 95 percent of population

2.5th. percentile   50th. percentile   97.5th. percentile

Souce:
The Measure of Man,
by Henry Dreyfuss,
Whitney Library of Design,
1967

	2.5th. percentile	50th. percentile	97.5th. percentile
Weight	58 kg	74 kg	95 kg
Span	1663	1798	1945

The plaintiff's attorney may bring charges of negligence against the designer.

To protect themselves in a product liability trial, engineers must use state-of-the-art design procedures during the design process. They must keep records of all calculations and methods used during the design process. Safety considerations must be included in the criteria for all design solutions.

The designer must also foresee other ways people could use the product. If a person uses a shop vacuum to remove a gasoline spill, is the designer responsible when the vacuum catches fire? The courts can decide that a design is poor if the engineer did not foresee improper use of the product.

It is imperative that you evaluate all of your alternative solutions against safety considerations. Reject or modify any unsafe elements of your design at this stage in the design process.

Figure 3-4, cont.
Anthropometric data

Standing adult female - including 95 percent of population

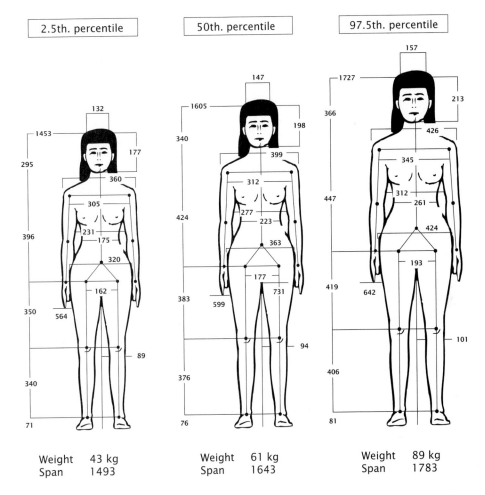

| Weight | 43 kg |
| Span | 1493 |

| Weight | 61 kg |
| Span | 1643 |

| Weight | 89 kg |
| Span | 1783 |

Economic and Market Analysis The net result or purpose of most engineering designs is to produce a product that generates a profit for the company. Obviously, each alternative design has to be evaluated against criteria such as sales features, potential market, cost of manufacturing, advertising, and so on. Large companies often conduct marketing surveys to obtain a measure of what the public will buy. These surveys may be conducted by telephone interviews with randomly selected people, or they may be personal interviews conducted with potential users of a product. Our society is based on economics and competition. Many good ideas never get into production because the manufacturing costs exceed what people will pay for the product. *Market analysis* involves applying principles of probability and statistics to determine if the response of a selected group of people represents the opinion of society as a whole. Even with a good marketing survey, manufacturers never know for certain if a new product will sell.

Strength and Mechanical Analysis Engineering analysis of a preliminary design is largely devoted to analyzing its mechanical features. The engineer conducts *mechanical analysis* to answer questions such as, "Will the device or structure support the maximum loads that it will be subjected to?" You must also determine the effect of shocks and repetitive or dynamic loading over the life of the product. Many systems generate heat, so you need to determine if the design can dissipate all of the heat being generated during normal operation. *Thermal analysis* is an area important to the design of electronic equipment. Many pieces of electronic equipment fail prematurely due to inadequate heat transfer. For example, the early releases of Intel's Pentium microprocessor could not operate at their rated speed due to overheating. The production of this microcircuit was delayed while engineers figured out ways to dissipate the excess heat.

You need to perform strength calculations to determine whether the design alternative will be able to support the specified mechanical loads. As a mechanical system is subjected to applied loads, it will deform or deflect. The following example illustrates a simple mechanical analysis on a proposed diving board design. This board must be designed for a maximum load of 500 pounds applied to the end of the board. The end of the board is required to deflect less than 1 meter for this load.

EXAMPLE 3-2   ## Diving Board

You are designing a diving board that must support a maximum load (force) of 225 kg on the end of the board (500 lb, a very large person!). The board is 4 meters long and consists of a rectangular cross-section with a height $h = 2$ cm and width $w = 25$ cm. Assume that the board is made out of plastic with given mechanical properties. Determine the maximum deflection at the end of the board.

SOLUTION

This is a classic mechanics problem, in which you are asked to determine the deflection of a cantilever beam to a load applied at the end. From a standard mechanics handbook, the equation defining the deflection of a cantilever beam to a load applied at the end is

$$h = \frac{PL^3}{3EI}$$

where

h = deflection

P = applied load = 225 kg

L = length of beam = 4.0 meters

E = modulus of elasticity = 2 x 10⁹ Pa

I = moment of inertia, m⁴

For a rectangular cross-section, the formula for moment of inertia is

Figure 3-5
**Mechanical deflection
of a diving board**

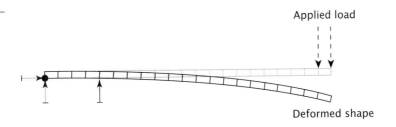

Applied load

Deformed shape

$$I = \frac{bh^3}{12}$$

$$= \frac{(.25).04^3}{12}$$

$$= 1.33333\,(10)\,\text{m}^4$$

Therefore, the maximum deflection of the board is

$$h = 1.8\text{m}$$

A deflection of 1.8 meters is nearly twice the allowable deflection specified for this design. To avoid damaging the board, you must modify the design to reduce the deflection for the applied load.

. . . .

You can evaluate each alternative design with a similar analysis, and you will incorporate the results into the decision phase of the design process. Example 3-2 consists of a very simple rectangular cross-section, and you can perform the analysis using basic algebraic relationships. Had the geometry of the proposed diving board design been any more complex than a rectangular cross-section, you would not have been able to calculate the deflection with such a simple analysis. Unfortunately, most design alternatives consist of complex geometry, and you need to use more complex methods to do the analysis.

Fortunately, modern computer tools allow the engineer to simulate and predict the response of a complex design solution to various mechanical operating conditions. With a proper computer model, you can change important parameters such as the length, thickness, or material of the board and observe the effect on the deflection. A computer simulation also allows for the analysis of geometries more complex than a simple rectangular cross-section.

Simple designs, consisting of a regular uniform geometry, can be analyzed using fundamental mathematical relationships covered in your engineering courses. Most real-life solutions, however, require the use of modern computer-aided engineering tools to generate an accurate analysis solution. One computer tool that allows an engineer to create an electronic model of a complex design solution and ask "what if" questions is the *finite element analysis* (FEA) method of analysis. The basic principle of the FEA is to take a complex object and divide it into simple regions, called *elements*. By selecting a simple geometry for each element, such as a triangle or quadrilateral, the mathematics are such that a solution can be found for the temperature, stress, strain, motion, and so on for that

Figure 3-6
**Finite element model
of a three-dimensional
part**

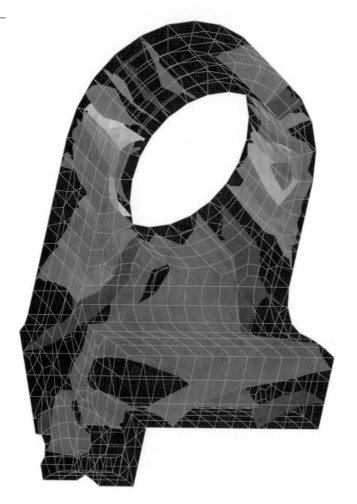

element. This technique is used to model the mechanical behavior of complex three-dimensional geometries. A typical finite element model is shown in Figure 3-6.

Used properly, FEA can greatly reduce the time and money spent building and testing prototypes, then redesigning and re-testing them to correct faults. FEA is used extensively in civil, aeronautical, and mechanical engineering to perform structural and thermal analyses, such as aerodynamic effects on aircraft wings, heating of engine parts, and stress on bridge trusses or building foundations.

The Decision Process   After analyzing your alternative solutions, you need to decide and document which design solution is the best. You will refine and develop the best solution in more detail during the later stages of the design process. At this stage, to evaluate each solution objectively against the stated design criteria or requirements, you need a quantitative basis for judging and evaluating each design alternative. One widely used method to formalize the decision-making process is the *decision matrix*. The decision matrix is a mathematical tool you can use to derive a number that specifies and justifies the best decision.

The first step in creating a decision matrix is for the design team to rank, in order of importance, the desirable attributes or criteria for the design solution. These attributes can include factors such as safety, manufacturing considerations, the ease of fabrication and assembly, cost, portability, and compliance with government regulations. You then assign to each attribute or criteria a *value factor* related to the relative importance of that attribute. For example, suppose you decide that safety is twice as important to the success of your design as cost. You would assign a value factor of 20 for safety and a value factor of 10 for cost. You assign value factors on a basis of 0 to 100, representing relative importance of each criterion to the decision.

Next you evaluate each design alternative against the stated criteria. A *rating factor* is assigned to each solution, based on how well that solution satisfies the given criterion. The rating factor is on a scale of 0 to 10, with 10 representing a solution that satisfies the given criterion the best. To make an accurate evaluation, you need as much information as possible. Unfortunately, engineers seldom have enough information to make a "perfect" evaluation. If you have done the analysis phase of the design process properly, those results can provide a basis for evaluation. Computer models and prototypes can also yield valuable information to assist in the decision phase. In most cases you must use engineering judgment, and the decision is subjective. The following example illustrates the use of a decision matrix in deciding the best alternative design for a can crusher.

EXAMPLE 3-3

Aluminum Can Crusher

Design a simple device to crush aluminum cans.

SOLUTION

A student design team proposes four solutions to the problem. They develop six criteria that are important to a successful design. The student team agrees that the most important criteria (or desirable attributes) of the design and assigned weights are

- Safety: 30 percent (30 points)
- Ease of use: 20 percent (20 points)
- Portability: 20 percent (20 points)
- Durability and strength: 10 percent (10 points)
- Use of standard parts: 10 percent (10 points)
- Cost: 10 percent (10 points)

This team also proposes four alternative solutions to this problem, which are illustrated in Figure 3-7:

1. A spring-loaded crusher
2. A foot-operated device
3. A gravity-powered dead weight crusher
4. An arm-powered lever arm crusher

Figure 3-7
**Alternative designs
of an aluminum can
crusher**

DESIGN IDEA 1

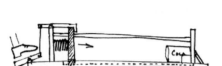

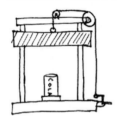

- SPRING LOADED CAN CRUSHER
- FOOT OPERATED TRIGGER
- CRUSHING PLATE REQUIRES
  LOWER GUIDE TRACK AND
  UPPER GUIDE BAR

DESIGN IDEA 2

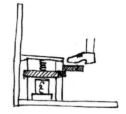

- FOOT OPERATED CAN CRUSHER
- CRUSHING PLATE REQUIRES
  TWO GUIDE TRACKS
- SPRING RETURNS CRUSHING
  PLATE TO STATIC POSITION

DESIGN IDEA 3

- GRAVITY CAN CRUSHER
  (USING POTENTIAL ENERGY)
- CRUSHING PLATE REQUIRES
  TWO GUIDE BARS
- MUST BE RELOADED BY
  PULLING ON CORD
- FINGER TRIGGER

DESIGN IDEA 4

- ARM POWERED CAN
  CRUSHER
- LEVER ACTION CRUSHES
  CAN
- SINGLE GUIDE TRACK
  FOR CRUSHING PLATE

Table 3-1   Decision matrix for evaluating alternative can crusher designs

Criteria	Weight(%)	Design 1	Design 2	Design 3	Design 4
Safety R x Weight	30	2 60	9 270	2 60	9 270
Ease of Use R x Weight	20	8 160	9 180	6 120	9 180
Portability R x Weight	20	5 100	3 60	2 40	8 160
Durability R x Weight	10	8 80	8 80	6 60	8 80
Standard Parts R x Weight	10	7 70	7 70	8 80	8 80
Cost R x Weight	10	6 60	5 50	7 70	8 80
Total	100	530	710	430	850

After analyzing each solution against the six criteria, the team evaluates each design alternative. After assigning a rating factor to each design alternative for each of the specified criteria, the team multiplies the rating factor by the value factor. The product of the value and rating factors is then summed down the column for each design alternative. The total sum at the bottom of each column determines the best design alternative. The results of this decision matrix are illustrated in Table 3-1.

Although rating each design against the six stated criteria is subjective, the rating factor for each design alternative is assigned according to the consensus of the design team. The results of an analysis are used to evaluate and rate each design. The rating factor R is assigned according to the following scale:

Excellent   9-10
Good   7-8
Fair   5-6
Poor   3-4
Unsatisfactory   0-2

Design 4 was chosen the best design largely due to the rating assigned for safety, criterion 1. The team felt that the chances of human injury were negligible for this design. Since safety is the most important factor (30% of the total weight), the high safety rating for design 4 gives it the highest overall score (9×30, or 270).

. . . .

Try It   Form a design team for the design of a residential entrance/exit for a household pet. After generating at least three alternative solutions to this problem, the team should

1. Choose criteria.

2. Assign weighting factors.

3. Develop a rating scale.

4. Construct a decision matrix.

5. Determine the best design alternative.

The validity of a decision matrix depends on the agreement of the design team, the experience of the team members, and the information available for each design solution.

What If   Assume that portability is not an important issue in the design of a can crusher. As a design team, you should

1. Reassign weights.
2. Redo the decision matrix.
3. Determine the best design alternative.

Step 5: Test and Implement the Solution

The final phase of the design process is *implementation*, which refers to the testing, construction, and manufacturing of the solution to the design problem. You must consider several methods of implementation, such as prototyping and concurrent engineering, as well as distinct activities that occur during implementation, such as documenting the design solution and applying for patents.

Prototyping The first stage of testing and implementation of a new product, called *prototyping*, consists of building a *prototype* of the product—the first fully operational production of the complete design solution. A prototype is not fully tested and may not work or operate as intended. The purpose of the prototype is to test the design solution under real conditions. For example, a new aircraft design would first be tested as a scale model in a wind tunnel. Wind tunnel tests would generate information to be used in constructing a full-size prototype of the aircraft. Test pilots then fly the prototype extensively under real conditions. Only after testing under all expected and unusual operating conditions are the prototypes brought into full production.

Concurrent Engineering Traditional design practices are primarily *serial* or *sequential*: Each step in the process is completed in order or sequence only after the previous steps have been completed. The implementation of the design occurs after a prototype or model is created from engineering drawings. A machinist working from the engineering drawings generated by a drafter or engineer, made the prototype. Only after creating a prototype of the design would the engineer discover that a hole was too small, parts didn't mate properly, or a handgrip was misplaced. The part would have to be redesigned and the process completed until a satisfactory solution was reached.

In the competitive manufacturing climate of the 1990s, the serial practice of design has proven inadequate. In a matter of months, a manufacturer may find that factors such as markets, material prices and technology, and government regulations and tax laws may have changed. This competitive environment requires a company to design high-quality products faster, better, and less expensively than their competitors. One solution to the traditional design paradigm is concurrent engineering.

Concurrent engineering is the ability to implement parallel design, analysis, and manufacturing processes. Concurrent engineering is only possible through the application of modern computer-aided design, analysis, and manufacturing software. A designer starts with an idea of a new product and uses the CADD software to create a preliminary design. With the appropriate software, the preliminary design can also be analyzed for factors such as strength, stresses, and weight as the design is being created. Using the results of this computer-generated analysis, the designer then makes any necessary modifications and reanalyzes the computer model. An engineer designing a bicycle frame would use concurrent engineering to minimize the weight and maximize the supported loads in a new frame design. The engineer would first create a design and model the physical behavior of the frame on the computer before actually manufacturing the frame.

The final stage in concurrent engineering is called *rapid prototyping* or sometimes called "art to part." As shown in Figure 3-8, the three-dimensional computer model of the finished design is used with *computer-aided manufacturing (CAM)* software to drive appropriate machinery to physically create the part. The entire design cycle therefore becomes nearly paperless. Engineers can go from design to prototype in a matter of days, instead of weeks or months as with earlier serial design practices. Since design is an iterative process, concurrent engineering significantly shortens the time between iterations. A product can therefore get to market much quicker, at a lower cost, and with a higher quality.

Figure 3-8
Rapid prototyping

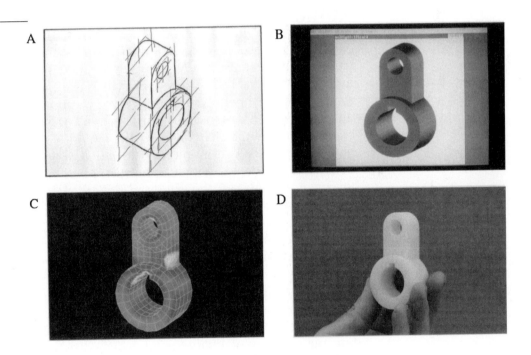

Try It

If a pair of swinging doors is chosen as the best alternative solution for the design of a residential entrance/exit for a household pet, describe the prototyping and concurrent engineering implementation.

Documenting the Solution   Once you've created a new product, system, or machine, your job as a design engineer is not finished. One of the most important activities you must do as you implement the design is *documenting* your work, clearly communicating the solution to your design problem so someone else can understand what you've created. Usually this consists of a design or technical report. Communicating the solution to a design problem through language, both written and oral, is a vital part of the implementation phase. Many people you will be communicating with do not have technical training and competence. They may be the general public, government officials, or business leaders. Successful engineers must possess more than just technical skills. The ability to communicate and sell a design solution to others is also a critical skill.

You can use graphs, charts, and other visual materials to summarize the solution process and present your work to others. Multimedia techniques, including overhead transparencies, slides, sounds, video, and computer-generated animations, are often used to clearly communicate the solution to a design problem.

Applying for Patents   If you develop an original and novel solution to a design problem, part of the implementation phase may include applying for a patent on your solution. A patent will not protect you from someone else copying your solution, but it does give you specific rights to make and sell your design for a specified period. A patent is an agreement made between you—the designer or inventor—and the U.S. government. Through a patent document you agree to make public all the details and technology of your invention. You agree to provide an *invention disclosure*, which provides enough details to allow anyone to build a working model of your invention. Most large libraries now have files of issued patents, which are available for anyone to see. These can be a good source of ideas for engineering design solutions. In return for making your invention or design solution public, the U.S. Patent Office grants you the exclusive right to make and sell your invention for a specified period of time.

Pursuing a patent is not a trivial process and may take a long time, costing hundreds or even thousands of dollars. Before considering a patent you should have a general understanding of patent requirements and what can be patented. Ideas by themselves cannot be patented. To obtain a patent, you must prove that your idea can be applied to produce a "new and useful process, machine, manufacture, or composition of matter, or any new and useful improvement thereof." These categories include just about everything made by people and the processes for making them.

Most engineering design problems fall into the patentable categories of utility patents or design patents. All mechanical and electrical devices fall into the category of *utility patent*, which is granted for 17 years. At the end of the patent period, your protection expires, and anyone can copy, manufacture, and sell your invention without giving you credit or payment. A *design patent* is granted to protect the styling or ornamental features of a design. A design patent is only granted for the appearance of an item, not for how it works or is made. For example, if you invent a telephone that looks like a shoe, you might apply for a design patent. The design patent would be granted on the appearance of the phone, not on the electronic and mechanical workings of the phone. Design patents are granted for 3-1/2, 7, or 14 years, depending on the patent fee paid. The fees range from $200 to $600.

Patents are only granted to the inventor of a device. However, the inventor can assign the rights to the patent to another party. If you develop an invention while working as an engineer for a company, you will probably be required to assign the patent rights to that invention to your employer.

Once a patent is granted, there is no guarantee that someone else will not try to copy the invention. The U.S. Patent Office does not enforce patent rights. It is the responsibility of the patent holder or a patent attorney to police the patent and make sure no one else copies it while it is in effect. Since a patent makes all information about your design public, some people choose not to pursue a patent, but rather keep the details of the invention secret. If no one else learns how the invention works, you will have protection until another inventor figures it out. For example, the formulas for Coca-Cola™ and Silly Putty™ have never been patented, and the secrets are only known by selected company officials.

To apply for a patent, you need to prepare and include the following items:

- A written document clearly describing your invention and stating that you are the original inventor. Enough information must be provided so that someone else can make your invention from the information you provide. You must also make claims about your invention which describe the features which distinguish it from already patentable material.
- Engineering drawings that follow the format documented in *Guide for Patent Draftsmen*, which is available from the U.S. Patent Office.
- The filing fee. This is a basic fee of at least $150 that must accompany the patent application. If the patent is granted, you will be charged an additional patent issue fee. The total charges for obtaining a patent can be hundreds of dollars.

A patent is granted only after an extensive review process of the U.S. Patent Office. The office will first search the nearly 5 million existing patents to determine whether your design has been previously patented or infringes on an existing patent. This process can take several years and be very expensive. Many inventors employ patent attorneys or agents to conduct a preliminary patent search. Most large libraries have records of all the patents filed with the U.S. Patent Office. This information is also available on a CD-ROM database at many libraries. You can look at this database and read the patent applications filed under the same product category as yours. This will give you a good idea how an application is written and might help you improve your own design. Before spending more time and money pursuing a patent, it is a good idea to find out if someone else has already patented your invention.

Testing and Verification   Testing and verification are important parts of the design process. At all steps in the process, you may find that your potential solution is flawed and have to back up to a previous step to get a workable solution. Without proper testing at all stages in the process, you may find yourself making costly mistakes later as the following two examples demonstrate.

EXAMPLE 3-4 ■ **The Mars Observer**

The case of the Mars Observer vividly illustrates the importance of proper testing and verification before the production of a design solution.

On August 21, 1993, the Mars Observer satellite approached Mars and prepared for a descent to place it into orbit around the red planet. The spacecraft's computer began a series of commands to pressurize the fuel tanks in preparation for the descent. When designing the spacecraft, NASA engineers made the decision to protect the delicate transmitter tubes from shock by shutting the transmitter down for a few minutes while the tanks were being pressurized. As planned, the computer shut down the craft's transmitter for the 14 minutes required to pressurize the fuel tanks, breaking all links with the controllers 200 million miles away on the earth. When the 14 minutes were up, the mission controllers anxiously waited to re-establish communication with the spacecraft. Pointing the Deep Space Network's huge dish antennas toward Mars, the mission controllers heard only silence. For the next several hours, the control center team began sending new commands to Mars, hoping that something would turn on the transmitter. All around the world, the huge antennas strained to hear something from the satellite. Several days later, NASA publicly admitted what many scientists and engineers feared from the beginning: The Mars Observer was destroyed or lost for eternity.

During the initial design of the Mars Observer, a decision was made to save $375,000, the cost to test the radio transmitter tubes for shock. Rather than test the radio transmitter tubes for shock, the engineers decided simply to turn the transmitter off during the bumpy 14 minutes while the Observer's propellant tanks were being pressurized.

Four months later, a Mission Failure Investigation Board released a report trying to piece together the events that occurred during the 14 minutes of silence when the transmitter was deliberately turned off. Because the communication link with the spacecraft was shut down during the incident, nobody knows for certain the cause of failure. The official report suggested that the probable cause of failure was an explosion on board when the propellant tanks were pressurized.

The Mars Observer represents years of hard work by hundreds of scientists and engineers. The cost to U.S. taxpayers was nearly $1 billion. Because a decision was made to shut down the transmitters, vital information was lost. If the Mars Observer had been transmitting data during the failure, the cause could have been determined with a high degree of certainty.

. . . .

EXAMPLE 3-5 ■ **The General Electric Rotary Refrigerator Compressor**

In 1986 General Electric decided to use a new rotary compressor in its domestic refrigerators. Rotary compressors were desirable to GE because they were much cheaper to build than the reciprocating piston compressors used in refrigerators since the 1920s. The designers of the new compressors overlooked the fact that rotating compressors require substantially more power than reciprocating compressors, and that their high speed makes them operate at higher temperatures.

The first rotary compressors were supposed to be tested for the assumed lifetime of the compressor before mass production. However, these tests were cut short because of pressure to get the new refrigerators to market quickly. Even though experienced technicians reported signs of overheating and wear in the compressors, GE management went ahead with producing and marketing the refrigerators.

The new refrigerators sold well for the first year they were introduced. Then consumers began reporting significant compressor failures. GE was forced to replace more than a million rotary compressors with conventional reciprocating compressors at a cost of $450 million to the company.

The cases of the Mars Observer and General Electric refrigerator are costly examples of why new designs need to be thoroughly tested under all potential operating conditions before the design is fully implemented. Testing is often costly, but pays big returns later in the process of designing a solution.

. . . .

3-3 COMMUNICATION AND THE DESIGN PROCESS

Communication is a vital part of every step in the design process. The ideas you generate and the work you do in solving the design problem must always be communicated to someone else. Therefore, a significant amount of your time as an engineer will be spent selling and explaining your ideas to either a supervisor or a customer. Productive engineers generate many types of technical publications, including reports, abstracts, instructions, memos, and letters. Besides written communication, you can also communicate graphically, through engineering drawings, and orally.

Engineering Drawings

Probably the most important type of documentation for engineering design problems is *graphical specifications,* or *engineering drawings.* Engineering drawings are a way to communicate or show graphically the exact size and shape of your design solution. You use engineering drawings to show someone else what your idea looks like and exactly how to produce it. Not only is the size and shape of your design shown in engineering drawings, but also information is given to show how the pieces fit together and how it operates. Verbal communication, while important, does not adequately convey a concept that is essentially visual.

During the implementation stage of the design, various craftspeople will use engineering drawings to manufacture and assemble your design. Therefore, they must be drawn clearly and according to standards and conventions accepted by the technical team. As an engineering student, you will be introduced to accepted conventions and standards in technical drawing through a course in *engineering graphics.* Engineers are generally not expected to be proficient drafters but are required to understand and use principles of technical drafting. A set of engineering drawings, produced by either yourself or a drafter, give enough unambiguous information for a machinist or craftsperson to make the design product. The machinist must know such things as what the finished product looks like, what the dimensions and tolerances are, what materials will be used, how the product is assembled, and how it is finished. Proper engineering drawings allow a shop to produce the same part or product every time it is made, no matter who is making it. Engineering drawings may number in the hundreds for a complex machine such as an automobile engine. These drawings include detail drawings and assembly drawings. *Detail drawings* show the exact sizes and locations of each part and piece in the engine. A detail drawing specifies all pertinent information necessary to make a single part. This includes shapes, all dimensions, locations of holes and other features, materials, and identifying part numbers. A detail drawing is laid out following the principles of *orthographic projection*. Orthographic projection is a method of describing a three-dimensional object on a two-dimensional piece of paper. In orthographic projection an imaginary box is placed around the object, with each side of the box representing a different view of the object placed at right angles to the other views. Each view is drawn on the same sheet of paper. An example of an orthographic detail drawing is shown in Figure 3-9.

Figure 3-9
Detail drawing

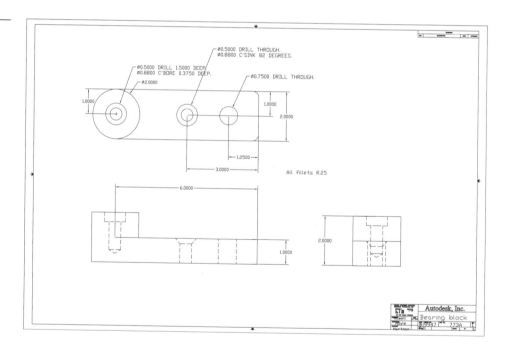

An *assembly drawing* illustrates how the individual parts fit together to make up the complete product. Both orthographic views and pictorial style drawings are included in an assembly drawing. The part numbers on an assembly drawing refer to part numbers on the associated detail drawings. A parts list with a complete bill of materials should also be included. Figure 3-10 shows an exploded assembly of a vise, showing how the individual parts fit together. More thorough information on both detail and assembly drawings will be covered in your engineering graphics course.

Figure 3-10
Assembly drawing

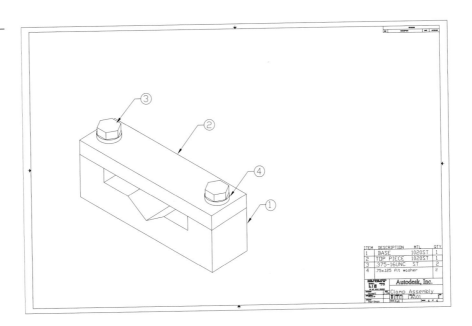

Written Communication

Although a good set of engineering drawings can communicate immense amounts of information about a design solution, written communication such as memos and technical reports are just as important.

Memos   Memorandums, or *memos*, are a common form of written correspondence used by engineers to convey information to people such as supervisors, fellow workers, or a design team. Unlike a technical report, a memo is brief, usually no more than one or two pages. The memo is a convenient way to document your progress and keep a permanent record of your work. Often memos are archived in a company's central file and so provide a permanent history of transactions and work done by the employees.

A memo may be written to one person, but more commonly it is sent to a *distribution list,* which is a list of people within the organization who have an interest in what you are saying. These people get a copy of the correspondence. Memos are often read by many more people than those on the original distribution list.

Most companies use a standard format for a memo. An example is shown in Figure 3-11. The memo header generally includes lines telling who the memo is to, who it is from, the date it was written, and the subject. The subject line functions as the title for the memo and is a brief synopsis or summary of the memo. Many companies have a standard memo template that gives the company name at the top of the memo.

Technical Reports   The *technical report* is a longer, more complete record of everything you did to solve a design problem. The basic rule for technical writing is the same as any other type of communication: The writing should be direct, clear, and readable for the intended audience. The first step in any type of writing is to know your intended audience. As a student, you write technical reports for teachers. Each teacher is an

Figure 3-11
**Example of
memo style**

INTEROFFICE MEMORANDUM

To: Stu Dent, Thermal Analysis Branch
CC: Marty Martin, Director of Engineering
From: Beverly Brown, Branch Manager
Date: 10/18/1995
Subject: Project Review Meeting for LX-1 Design Modifications

---

The LX-1 design team will meet on November 2, 1995 for our second project review meeting in room 201 of the engineering building. Please be prepared to discuss the results of the stress analysis you performed on the LX-1 mounting bracket. At that meeting we will decide whether it is necessary to proceed with the proposed modifications to the mounting bracket. As you know, the modifications will delay the final production of the LX-1 by at least 3 months.

---

expert who understands the field. At school you can assume that a teacher is a captive audience and will read all of your report, no matter how time-consuming and uninteresting the job. As a practicing engineer, however, you are the expert with a problem solution to present to your readers. The people reading your writing are very busy and probably reading numerous letters, memos, and reports in a day. They want to learn and understand the main points of the document and are unwilling to wade through irrelevant and vague information to attempt to understand what you are saying. The audience for professional writing needs the information you are communicating presented as clearly and unambiguously as possible.

The types of reports you will write as an engineer will be varied, so we cannot present a specific outline or format that you can apply to every report. However, the following information should, in general, appear in all reports:

- Cover page, stating title of project, company name, your name, and date
- Abstract, giving a brief overview and summary of the work performed
- Table of contents
- Body of report, which elaborates on the problem solved, presents background material, procedure used, and results
- Conclusions and recommendations, which summarize the results and significance of your work
- Appendices, which give details for the reader who wants to know everything about your work

Since technical readers are looking for specific information, they will most likely not read every word you write. Therefore, it is important that every technical report begin with a brief abstract that summarizes the purpose and results of the effort you are reporting before giving any details. Many people will only read the introduction or abstract of a report and then jump right to the conclusions. People who have an interest in the details of your conclusion will read the body of the report. The body of the report elaborates on the problem definition, gives necessary background material, discusses the strategy used to solve the problem, and presents the results of any analysis or research performed. You save the intimate details of your work for the appendices. Detailed derivations of equations, tables of data, and other details are given in the appendices to avoid interrupting the flow of your report. The conclusions and recommendations summarize the work you did and highlight important results. This is the section that sells your idea to the reader. It should be brief and highlight the important points of your solution.

Oral Communication

At different stages during the design process, you will most likely be called on to give oral progress reports to other members of the design team, your supervisor, or sales and marketing people. Your success as an engineer may depend on your ability to give a clear and effective oral presentation. The principles and objectives of an oral presentation are the same as for a written report: You want to communicate information and convince the audience. The format and methods you use, however, are very different. A written report is designed to communicate the details and can be studied and reread by the reader. An oral presentation, however, presents only the highlights or important points. You must communicate the information to the audience concisely over a short time. Therefore, your presentation must be simple. A good oral presentation does not give all the details of a project; the listener can go to the written report for that information. You do not have the time to include complicated equations, tables of data, or complicated graphs.

The most important element to a successful oral presentation is preparation. Obviously, you should be intimately familiar with the subject. You should also know exactly how much time is allotted for the presentation and practice until you can cover everything completely in the allotted time. As with all forms of communication, you need to know your audience. The level of the presentation must match the technical level of the audience. A presentation to nontechnical people will obviously be at different level than a presentation to a group of fellow engineers.

A good oral presentation involves both visual and auditory modes of communication. Having clear, concise, and interesting visual materials is as important as being an eloquent speaker. The quality of your graphical displays is usually more important than your speaking ability. Visual aids must be simple and uncluttered. Do not try to express too much information in a single visual display. If you clutter the visual with too many details, the main points will be lost to the audience. Several good computer programs are available to assist you in preparing high-quality visual aids.

Finally, have a good summary and conclusion to highlight the important points of your presentation. Reinforce the main points, and summarize them with a good closing visual aid. Never end your presentation with an apology, such as "Well, I don't really have any more to say."

3-4   SCHEDULING AND PLANNING A DESIGN PROJECT

Solving a design problem is much more complex than doing a homework problem in a physics class. As discussed previously, a complete solution requires several steps or tasks. A complex problem may require weeks or months to complete and involve various personnel. Obviously, a design problem solution requires that you work out a plan for the scheduling and sequence of tasks. Your objective is to develop a plan that has each task accomplished before its result is needed and makes use of all the personnel all of the time. You also need to schedule design reviews at the end of each phase in the process.

A common technique used to develop a schedule for a simple project is the *Gantt chart*. The Gantt chart is a bar chart showing each task plotted against a time scale (the units are usually dates, weeks, or months). The total personnel requirements for each time unit are also plotted, in addition to the schedule of the design reviews. A basic Gantt chart is shown in Figure 3-12. As work on the project is completed, you can color in the bars, allowing you to see the timeliness of the project and keep the project on schedule.

Figure 3-12
Gantt chart for scheduling a student design project

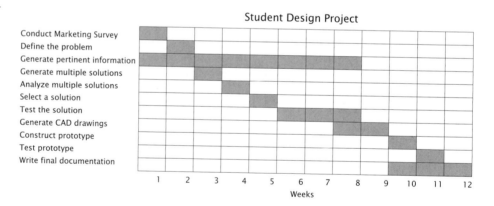

This method is useful for small projects, such as the ones you will do as an engineering student, but becomes too cumbersome for a more complex project with many tasks. It also does not indicate which parts of the project can fall behind schedule and not hinder the progress of the entire project. Two methods you can use in developing an efficient schedule for a complex project are the *critical path method (CPM)* and *program evaluation and review technique (PERT)*. These methods take into account which parts of the project are critical to the progress of the project. A complete discussion of these methods is beyond the scope of this text but may be found in many other sources.

SUMMARY

This chapter discusses the application of the general five-step problem-solving process to design problems. The solution to a design problem is usually a system or device that meets a human need. Design problems differ from other engineering problems because they generally have several correct solutions. The solution path for a design problem most likely involves iteration or backtracking. Chapter 4 will present a sample student design project to illustrate these steps. The five steps are:

1. Define the problem.
2. Gather pertinent information.
3. Generate multiple solutions.
4. Analyze and select a solution.
5. Test and implement the solution.

Key Words

anthropometric data
assembly drawing
brainstorming
computer-aided manufacturing
  (CAM)
concurrent engineering
contingent process
creativity
criteria for success
critical path method (CPM)
decision matrix
design patent
design team
detail drawing
distribution list
documenting
element
engineering drawing
engineering graphics
ergonomics
finite element method (FEM)
functional analysis
Gantt chart
graphical specification

implementation
index
invention disclosure
liability
market analysis
mechanical analysis
memo
orthographic projection
program evaluation and review
  technique (PERT)
prototype
prototyping
rapid prototyping
rating factor
safety
sequential
serial
sketchstorming
synergism
technical report
thermal analysis
utility patent
value factor

Exercises

1. Write a complete problem definition statement, including a list of criteria for success, from the following general statement of need: Design a better snowboard binding.

2. Write a complete problem definition statement, including a list of criteria for success, from the following general statement of need: Design an improved bicycle rack for cars.

3. Write a complete problem definition statement, including a list of criteria for success, from the following general statement of need: Design a device to help arthritic patients open child proof medicine bottles.

4. Write a complete problem definition statement, including a list of criteria for success, from the following general statement of need: Design a baby stroller for joggers.

5. Generate preliminary ideas or alternative designs for the problem definition statements you developed in exercises 1, 2, and 3. These designs should be subject to the constraints you have imposed.

6. In groups selected by your instructor, brainstorm and generate at least five different systems for separating the following refuse components for recycling: ferrous materials, glass, plastic, and aluminum. Use sketches to communicate your ideas.

7. In groups selected by your instructor, brainstorm and generate at least 5 ideas for a better mousetrap using the following problem definition: Design a mousetrap that protects humans from exposure to viral and bacterial contamination carried by the mouse.

8. Identify and list some improvements to your current classroom. Record these ideas with sketches.

9. Using the information resources available through your library, investigate the problems associated with battery-powered automobiles. Create a preliminary design of a battery-powered automobile that addresses the shortcomings of existing electric cars.

10. Using the information resources available through your library, investigate the problems inherent with the initial design of the Hubble space telescope. How could this design have been better tested before its implementation?

11. When an airplane touches down at landing, the wheels are nearly instantly brought up to a speed of about 200 mph due to the friction between the tire and the runway. Generate some preliminary designs for a landing wheel spinner that spins the wheel prior to touchdown, thereby reducing the friction and heat generated in the tire.

12. Develop a problem definition statement and criteria for success for the design of the following:

    (a) A method to avoid locking keys inside an automobile

    (b) An indicator to tell when an automobile is not in motion

13. Choose an idea for a new automobile accessory and develop the problem definition statement and criteria for success.

14. Choose five design problem areas (such as home entertainment, fitness, nutrition, and so on) and indicate at least three major sources of information for each area.

15. Form a design team and brainstorm to create alternative solution ideas for the design of the following:

    (a) A method to avoid locking keys inside an automobile

    (b) An indicator to tell when an automobile is not in motion

16. Form a design team, and apply the decision matrix analysis method to the design of an automobile accessory the team selects.

17. Create a Gantt chart for the design of an automobile accessory of your choice.

4 A Student Design Project

Hard Disk Drive
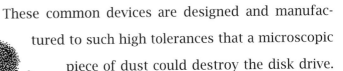

The miniature hard disk drive in a portable computer can store information equivalent to thousands of books on a device that fits in your shirt pocket. The information is stored electronically on a rapidly spinning magnetic disk less than three inches in diameter. A tiny recording head moves precisely across the disk surface at a height of less than 10^{-6} mm without touching the surface. These common devices are designed and manufactured to such high tolerances that a microscopic piece of dust could destroy the disk drive. This type of engineering accomplishment requires the specialized talents of the members of a team of engineers and scientists. Learning to work in project teams is a crucial part of engineering education. This chapter presents the design process used by a team of engineering students.

INTRODUCTION

I n Chapter 2 you learned a five-step problem-solving process for closed-ended analysis problems. This process was expanded and modified in Chapter 3 to solve open-ended design problems. To illustrate each step in the problem-solving process and help you make a distinction between analysis and design problems, this chapter documents each step of a design project for beginning engineering students.

4-1 A SAMPLE STUDENT DESIGN PROJECT

This chapter examines a freshman student design project from start to finish to illustrate all aspects of the design process.

Application 1   **WATER BALLOON/SNOWBALL LAUNCHER**

1. Define the Problem

This project begins by identifying a need presented by the vice president of sales and marketing for the PayNPlay Toy Company. The toy company conducted a marketing survey and determined a potential market for a new product, a water balloon/snowball launcher. A statement of need was presented to the director of engineering, represented by the design team. The need was deliberately vague and general, so the first task of the student design team was to produce a clearly defined problem definition statement. The memo shown in Figure 4-1 was presented to the student design team, documenting the need to design and produce a new product.

Figure 4-1
Statement of need for design projec

INTEROFFICE MEMORANDUM

To: Stu Dent, Director of Engineering
From: Sheryl Barrel, VP of Sales and Marketing
Date: March 22, 1994
Subject: New Product of PayNPlay Toys

The Market Research Team at PayNPlay Toys recently conducted a study to determine the market potential for new toys targeted to the 10- to 18-year-old child. The results of this study indicate an interest in a new product: a water balloon or snowball launcher.

You and the engineering staff are requested to begin preliminary design of this product. This new toy should be capable of launching and propelling a one-pound water balloon or snowball. Factors to be considered in the preliminary design are: cost, safety, and ease of manufacturing.

I expect to be regularly updated on the status of this project through written progress reports and regularly scheduled design review meetings.

Figure 4-2
Problem definition statement

INTEROFFICE MEMORANDUM

To: Sheryl Barrel, VP of Sales and Marketing
From: Kevin Reed, Brett Hurtienne, Engineering Department
Date: April 6, 1994
Subject: New Product of PayNPlay Toys, Snowball/Water Balloon Launcher

---

The engineering staff has examined the need for a snowball/water balloon launcher for children that you presented in your memo of March 22, 1994 (re: New Product for PayNPlay Toys). After consulting the engineering staff, we have determined the following criteria are important for a successful design:

1. The launcher must have launch capability of at least 100 feet for a one-pound snowball or water balloon projectile.
2. The launcher must be safe to operate and meet all state and federal safety regulations.
3. The launcher must be operable by a single child.
4. The launcher must have a minimum manufacturing and production cost.
5. The launcher must be portable and easy to set up.
6. The launcher will require no external sources of power.

In the initial states of the design process, we are actively pursuing a device that will utilize either rubber bands, coil springs, wound springs, or a combination of some or all of these. These methods of propulsion are both safe and easy to use. The designs that we are considering will utilize a catapult mechanism, elastic potential in rubber, or a fixed length of rope in a circular motion to launch the projectile. We hope to have a finalized design by May 1, 1994.

In response to the memo from the vice president of sales and marketing, the student design team began to develop a clearer definition of the problem. They decided what criteria were important in a successful design. This formed the basis for the problem definition recorded in the memo shown in Figure 4-2. After some discussion the design team developed the list of criteria for success: They decided that the device should be able to launch a 1-pound water balloon or snowball at least 100 feet, must be safe to operate, should be operable by a single child, must be portable, should be inexpensive to produce, and must operate without any external power. As the design process progressed, this team found that the first criterion (100-foot launch range) contradicted the second stated criterion, safety. Any device capable of launching a 1-pound projectile for a distance of 100 feet is inherently unsafe. Later in the design process, the design team will need to refine and modify these criteria as more information becomes available.

2. Gathering Pertinent Information

Information gathering for this project consisted primarily of looking at existing products. The student design team discovered one such water balloon/snowball launcher constructed of rubber surgical tubing. The team also interviewed fellow students who had used the existing launcher and asked them questions about the features, price, and desirable characteristics. This existing solution formed the basis for one of their alternative solutions. The results of their information gathering are documented in the memo written to the VP of sales and marketing shown in Figure 4-3. Since the initial information gathering was not adequate to complete the solution to this problem, the team was forced to gather relevant information at all stages of the design process (such as material physical properties, costs, and manufacturing techniques).

Figure 4-3
Results of information gathering

INTEROFFICE MEMORANDUM

To: Sheryl Barrel, VP of Sales and Marketing
From: Kevin Reed, Brett Hurtienne, Engineering Department
Date: April 10, 1994
Subject: Survey of Existing Snowball/Water Balloon Launcher

The engineering staff has visited all the toy stores in our area and examined the existing snowball/water balloon launchers now on the market. We have found only one launcher is currently available. This launcher has a retail price of $24.95 and is a slingshot device using rubber surgical tubing to provide the motive power for the projectile. The launcher requires two people to hold each end of the surgical tubing plus a third person to actually launch the projectile.

We also interviewed students at Northern Arizona University who have used this launcher in water balloon fights. They expressed general satisfaction with the range of this launcher. They've been able to place water balloons into the student commons area from a distance of over 100 feet.

However, these same students expressed dissatisfaction with the existing launcher for several reasons:
1. The existing launcher requires three people to operate. Two people hold the ends of the surgical tubing. The third person loads the water balloon and stretches the elastic to launch the balloon.
2. The existing launcher is difficult to aim accurately.
3. The existing launcher is not safe. Several students reported that the people holding the ends of the elastic were accidentally hit by the balloon at close range.

Based on the results of our survey, we are confident that we can produce a launcher the same range but can be operated by a single person. We also feel that we can improve the aim of the existing launcher and reduce the risk of accidents.

3. Generate Multiple Solutions

The design team generated multiple solutions with a brainstorming session. They spent a short time (less than an hour) discussing potential methods of launching a water balloon or snowball. These ideas were recorded on paper with quick sketches. After the brainstorming session, the group rejected some of these ideas because they were impractical. The three best ideas were sketched using common annotation techniques. A memo was generated to document any special features and discuss the operation of each idea. These results are shown in Figures 4-4 and 4-5.

Figure 4-4
Preliminary solutions

INTEROFFICE MEMORANDUM

To: Sheryl Barrel, VP of Sales and Marketing
From: Kevin Reed, Brett Hurtienne, Engineering Department
Date: April 14, 1994
Subject: Results of Brainstorming on Water Balloon/Snowball
           Launcher Design Problem

The engineering team met to generate potential solutions to the problem of designing a water balloon/snowball launcher for PayNPlay Toys. The results of our brainstorming session generated nearly 30 possible devices that might meet the stated design constraints.

After much discussion, we refined this list down to three alternative solutions. The three ideas are shown in the attached sketches. At this point in the design process we are actively pursuing a device that will utilize either rubber bands, coil springs, wound springs, or a combination of some or all of the above. The designs that we are considering will utilize a catapult mechanism, elastic potential in rubber, or a fixed length of rope in a sling type arrangement to provide the angular momentum to launch the projectile.

---

4. Analyze and Select a Solution

The design group used a decision matrix to evaluate the top three alternative solutions against the stated criteria for success. The team selected the following criteria as important for a good solution:

1. Safe (most important with a value factor of 20)
2. Achieves a launch distance of 100 feet (value factor of 17)
3. Easy to use (value factor of 15)
4. Affordable for the intended market (value factor of 14)
5. Fun to use (value factor of 12)
6. Provides a high profit margin, retail cost to production cost (value factor of 11)
7. Uses standard, available parts (value factor of 10)
8. Durable (value factor of 9)
9. Portable (value factor of 8)

Figure 4-5
**Sketches of
preliminary solutions**

4/4/94

ALTERNATE SOLUTION 1 - SNOWBALL/WATER BALLOON LAUNCHER

- MECHANICAL

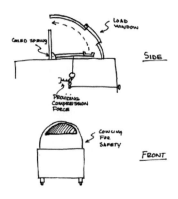

SIDE

FRONT

- ROLLING CHASSIS TO ALLOW AIMING AND MOBILITY
- COMPLETELY CONTAINED MECHANISM FOR SAFETY
- LOAD WINDOW ALLOWS ACCESS TO LAUNCHING ARM
- EXPENSIVE TO PRODUCE?

4/4/94

ALTERNATE SOLUTION 2 - SNOWBALL/WATERBALLOON LAUNCHER

- FIXED SLING SHOT

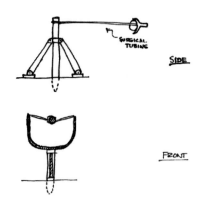

SIDE

FRONT

- EASILY ATTAINS LAUNCHING DISTANCE
- EASY TO USE AND SETUP
- SAFETY A PROBLEM?

4/4/94

ALTERNATE SOLUTION 3 - SNOWBALL/WATERBALLOON LAUNCHER

- BOLA STYLE

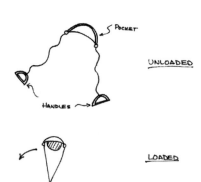

UNLOADED

LOADED

- SIMILAR TO OLYMPIC HAMMER THROW
- EASY TO USE
- CHEAP TO PRODUCE
- SAFE?

The decision matrix used to evaluate this design project is shown in Table 4-1.

Table 4-1   Decision matrix

Goal	Bola	Slingshot	Machine
Safe (20)	2(20) = 40	6(20) = 120	7(20) = 140
Achieves launch distance (17)	5(17) = 85	10(17) = 170	7(17) = 119
Easy to use (15)	3(15) = 45	8(15) = 120	9(15) = 135
Affordable (14)	10(14) = 140	9(14) = 126	2(14) = 28
Fun (12)	5(12) = 60	9(12) = 108	7(12) = 84
Provides high profit margin (11)	9(11) = 99	8(11) = 88	3(11) = 33
Uses standard parts (10)	9(10) = 90	8(10) = 80	6(10) = 60
Durable (9)	6(9) = 54	7(9) = 63	4(9) = 36
Portable (8)	10(8) = 80	7(8) = 56	3(8) = 24
Total	693	931	659

The decision matrix is an unbiased document that supports the design team's decision to carry out and refine a particular design alternative. The value factors assigned to each criteria were decided by the design team after much discussion. The factors were ranked in order of importance, and the numbers were assigned based on relative importance to a successful design. Safety was given the highest priority because of the potential for serious injury from a device that propels a 1-pound projectile for a distance of 100 feet. A toy such as this is inherently dangerous and must include proper warnings, although there are potential liability problems. The team next rated how well each alternative design achieved the stated criteria on a scale of 0 to 10, with 10 being the highest rating factor. The number in each cell of the decision matrix is the product of the rating factor times the value factor. Finally, each column was summed to give a total score for each design. Based on the assumptions and evaluations done by this team, the second design, the slingshot, has the highest score and is therefore the best alternative.

5. Test and Implement a Solution

Once the design team decided the slingshot approach best met the stated design criteria, they began to analyze and put into place the details for this solution. Calculations were performed to determine the dimensions and sizes of all components. In particular, they needed to determine the length of rubber surgical tubing needed to propel a 1-pound projectile. The team did not have the experience or technical background to do a detailed mathematical analysis of the path of the projectile. Because of this lack of engineering experience, the team had to use an empirical technique to determine the required length of rubber tubing. They conducted an experiment using different lengths of rubber tubing and arrived at an optimum unstretched length of 6 feet.

As the solution continued to be refined, more information was collected on construction techniques and materials. Vendors were contacted and prices for the components were established. These prices were used to construct the parts list and cost analysis shown in Table 4-2.

Table 4-2   Cost analysis of prototype

Part	Material	Unit	Cost/unit ($)	Amount	Total Cost
Wishbone	1018 tube 1/16" 6" x 6"	Feet	1.65	10 ft	16.50
Collar	1018 tube 1/16" 3" x 3"	Feet	1.47	4 ft	5.88
Base Plate	1018 tube 1/16" 12" x 3"	Feet	2.12	1.5 ft	3.68
Base Plate	1018 tube 1/16" 3" x 1.8"	Feet	3.00	.5 ft	1.50
Base Plate	1018 sheet 1/16"	Feet^2	.87	1 Ft^2	.87
Joining Arm	1018 tube 1/16" 6" x 2.9"	Feet	5.50	4 ft	22.00
Joining Arm	1018 Sheet 1/16"	Feet^2	.87	484 Ft^2	.44
Bolt	.6 x 10.8 in steel	Individual	.19	4	.76
Nut	.6 in nylon	Individual	.13	4	.52
Elastic Tubing	.5" diameter	Feet	.23	24	5.52
Plastic Cap	Injection molded ABS	Individual	2.00	2	4.00
Pocket	Injection molded ABS	Individual	9.00	1	9.00
Total					70.67

The students used a CAD program to produce a set of fully dimensioned detail drawings and an assembly drawing. A final report was written, summarizing all the work done, giving a detailed breakdown of the cost of construction, and describing how the solution was operated and assembled. An oral presentation was given to summarize and describe the solution to this design problem. Finally, the design team wrote their final documentation, as shown in Figure 4-7. This document included a set of instructions describing the setup and use of the snowball launcher.

Solving Analysis Problems in the Design Project

The design of a snowball launcher is an example of an open-ended design problem. To solve any design problem, you must also solve many intermediate closed-ended analysis problems along the way. Once you have determined the general form of the solution (an elastic band slingshot system in this case), you need to solve smaller analysis problems, such as the following before fully specifying the final solution:

- **Length of the elastic tubing.** The length of the elastic tubing is specified by the maximum launch distance and the mass of the snowball. Calculating this requires an in-depth knowledge of physics and calculus and is beyond the experience of a beginning engineering student. For this project the design team experimented with different lengths of tubing until they had a length that worked for the given projectile.

Figure 4-6
**Engineering drawings
of completed design**

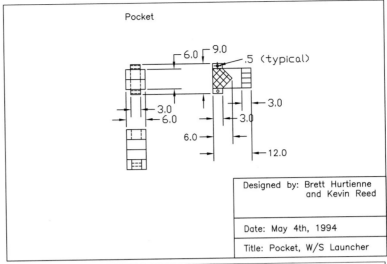

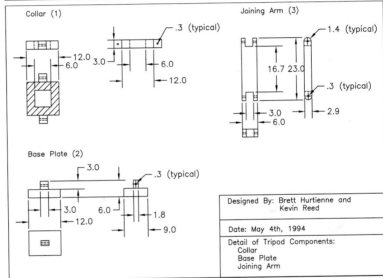

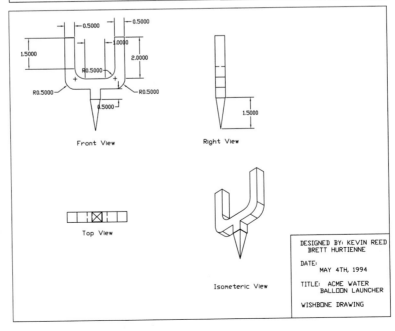

Figure 4-6, cont.
**Engineering drawings
of completed design**

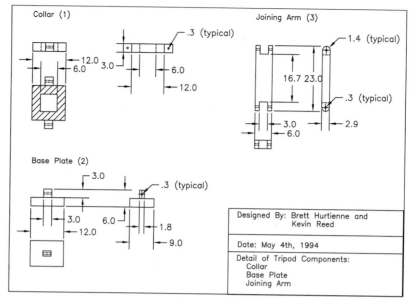

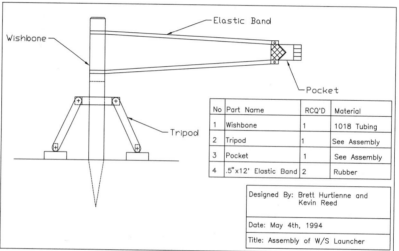

- **Size of the mechanical components.** The size of the mechanical components, such as the base plate, wishbone, collar, and joining arm, require a knowledge of mechanical design and strength of materials. A computer-generated finite element model would assist in determining the optimum size and shape of these components for a specified applied load. This type of analysis is beyond the expertise of beginning engineering students. These students specified a size that seemed right to them. To get an optimum design solution, the team would have to test and verify these sizes.

This design project is a *conceptual design.* A conceptual design does not have all the details fully worked out. Because this project was done by beginning students, the exact size of each component was not specified from detailed calculations or analysis. The next step in the design process would be to determine the optimum size of all mechanical components, which is an analysis problem.

Figure 4-7
**Final documentation
with instructions**

SET UP AND USE OF SNOWBALL LAUNCHER

Launcher Setup

1. Slide the wishbone (1) into the tripod (2) so that tripod is perpendicular to opening in the wishbone.
2. To drive the wishbone (1) into the ground, hold it vertically with tripod (2) on ground pointing in the direction you wish to fire. Then strike the wishbone with a hammer inside of the U-shaped portion directly above the center support.
3. The pocket (3) will be attached to the elastic bands (4) by the factory. Notice that one band threads through the upper hole on the pocket and one band threads through the lower hole on the pocket with equal lengths of elastic band on each side of the pocket holes.
4. With the pocket (3) opening facing toward the wishbone (1), attach one of the lower elastic bands (4) to each side of the wishbone by sliding the factory-installed loop over the wishbone posts. The bands should make a V shape, with the pocket at the point. Slide the loop down to 6 inches about the 90-degree turn of the wishbone tubing. Repeat this step with the upper tubing, but slide the loop 6 inches below the plastic cap on the top of the vertical wishbone posts.
5. The launcher is now ready to fire.

Launcher Use

1. With the elastic bands slack and the pocket pointing up, place the object in the pocket.
2. Grasp the pocket handle with both hands, making sure to keep the pocket at a slight up angle to keep the launch object in the pocket.
3. With the pocket held in a position behind and between the vertical wishbone uprights, begin to walk backward until sufficient tension is placed on the pocket to launch the object the desired distance. Again, the pocket should be at a slight up angle to keep the object in the pocket and to assure maximum firing distance.

Safety Notes

- MAKE SURE THAT NO PERSON OR OBJECT OBSTRUCTS THE PATH OF THE PROJECTILE (EXCEPT THE TARGET).
- MAKE SURE THAT 20 FEET IS ALLOWED IN FRONT AND TO THE SIDES OF THE WISHBONE TO ALLOW FOR EXTENSION AND RECOIL OF THE ELASTIC BANDS.

SUMMARY

In this chapter you followed a student design project from start to finish. This project began with a statement of need: A marketing survey from the vice president of marketing identified a market for a new toy snowball launcher. From this statement of need, a team of engineering students developed a list of criteria for success and a problem definition statement. These students gathered information related to the design of a snowball launcher capable of propelling a 1-pound snowball for a distance of 100 feet. The design team then generated alternative solutions to the problem and selected the best three for further development. They used a decision matrix to evaluate each alternative design against the stated criteria for success and determined the best solution. Finally, they produced a complete set of CAD drawings and a written report of the final solution.

Exercises

Use the method presented in this chapter to create a conceptual design of the following devices.

1. A portable bicycle rack for automobiles.
2. A bicycle rack to fit in the bed of a pickup truck.
3. A device to assist disabled people getting into a swimming pool.
4. A weatherproof luggage rack and container to carry a small package (up to 1 ft^3) on the rear of a bicycle.
5. A portable hunting blind.
6. A convertible or fold-down seating system for a cargo transport aircraft. This would be used when a cargo plane needed to also transport people.
7. A solar-powered oven.
8. A portable carrying case or system for holding cassette tapes or compact disks that can be used in an automobile.
9. A low chair or stool on rollers or wheels that can be used when gardening.
10. A portable seat back for bench-style stadium seats.
11. A simple built-in jack for raising automobiles.
12. An aluminum can crusher.
13. A ceiling-mounted rack for a classroom television set. The rack should be able to accommodate different-sized TVs.
14. A low-cost greenhouse for the home gardener.
15. Portable cold frames for the home gardener.
16. An inexpensive loft-bed system for dormitory rooms.
17. A system to recover rainwater from a gutter system to be used for watering gardens.
18. A solar-powered food dehydrator.
19. An antitheft system for automobiles.
20. Improved bicycle racks.
21. A car beverage or cellular phone holder.
22. A car CD storage unit.
23. A notebook PC carrying case.
24. A slug trap.
25. A portable diskette holder.

Index